Environmental Science and Engineering

Environmental Science

The protection of our environment is one of the most important challenges facing today's society. At the focus of efforts to solve environmental problems are strategies to determine the actual damage, to manage problems in a viable manner, and to provide technical protection. Similar to the companion subseries Environmental Engineering, Environmental Science reports the newest results of research. The subjects covered include: air pollution; water and soil pollution; renaturation of rivers; lakes and wet areas; biological ecological; and geochemical evaluation of larger regions undergoing rehabilitation; avoidance of environmental damage. The newest research results are presented in concise presentations written in easy to understand language, ready to be put into practice.

More information about this subseries at http://www.springer.com/series/3234

Gero Benckiser
Editor

Soil and Recycling Management in the Anthropocene Era

Springer

Editor
Gero Benckiser
Department of Applied Microbiology
Justus Liebig University (Retired)
Giessen, Hessen, Germany

ISSN 1863-5520 ISSN 1863-5539 (electronic)
Environmental Science and Engineering
ISSN 1431-6250 ISSN 2661-8222 (electronic)
Environmental Science
ISBN 978-3-030-51885-1 ISBN 978-3-030-51886-8 (eBook)
https://doi.org/10.1007/978-3-030-51886-8

This Springer imprint is published by the registered company Springer Nature Switzerland AG
The registered company address is: Gewerbestrasse 11, 6330 Cham, Switzerland

Foreword

Everywhere and invisible, vital and toxic: Justus Liebig had the idea in the 1840s and experimented with how to bring nitrogen from the air into the soil in a useful and profitable way. Fritz Haber and Carl Bosch succeeded in doing so at the beginning of the twentieth century, convert the nitrogen into ammonia and thus into nitrogen that can be used by plants. Liebig's drive was to avoid starvation in a growing world population that increasingly lived in cities. The corresponding increase in food was his antidote, while Georg Büchner, who also briefly stayed in Gießen, declared war on the palaces in the "Hessischer Landbote" and put the peasant revolution on the agenda. In the meantime, the "Green Revolution", which was initiated with the Haber–Bosch process, has won. Just a few years ago, the United Nations believed it could finally remove the problem of hunger from the list of the greatest concerns for mankind. It cannot be deleted because the processing of nitrogen and the consequential problems that this has caused have caused massive ecological damage which also threatens the food supply. Humanity has also reached planetary boundaries in this area of scientific and technical progress and must take stock.

The blessing and the curse already existed in the First World War, when ammonia was also used to produce tons of deadly explosives. A problem then became especially the expansion of the nitrogen quantities in the circuits, exacerbated by the NOx problem, the burning of fossil and renewable raw materials. Nitrate in the groundwater and nitrous oxide in the atmosphere is familiar to every layperson, the use of synthetic fertilizers has to be drastically restricted. And, there we are almost back to Liebig's initial problem: how to feed a still growing population, enriched by the existence of an important question, how to do it in a sustainable and healthy way. There are technical compensation options, national and European regulatory authorities exist, structures and organizational forms of agricultural production that have become anachronistic and are kept alive by strong but short-sighted lobby groups.

The nitrogen problem has long since reached planetary proportions and also requires transnational, multilateral interventions. It is at the crossroads of almost all environmental damage (water and air quality, greenhouse gases and species extinction, ozone) and affects almost all of the United Nations' sustainability goals.

Under the aegis of M. Sutton from the Center for Ecology & Hydrology (CEH) in Edinburgh, a comprehensive problem analysis was presented in 2019 on behalf of the International Nitrogen Management System (INMS), which culminated in an alarm call from the scientists: The report is framed as a manifesto for "Science-in-Action". By this, we mean applying the evidence from science and the experience of researchers engaged in policy development to develop a vision of how Sustainable Nitrogen Management must be a core theme for sustainable development. The report is aimed at stimulating discussion, raising awareness and catalyzing action by governments, business, civil society, scientists and citizens for a more sustainable Nitrogen Earth. We present options as opportunities for environment, health and economy. In identifying these opportunities, we point to benefits and limitations of particular actions, while avoiding recommendations for specific policies. These must be for governments to agree, locally, regionally, nationally and internationally, in the light of the emerging evidence".

May this book make a contribution.

Figure ES3: Nitrogen is almost everywhere relevant, but currently invisible in the UN Sustainable Development Goals (SDGs). The figure illustrates the multiple ways in which Sustainable Nitrogen Management can contribute to meeting the SDGs, highlighting the potential of an ambitious aspiration to Halve Nitrogen Waste globally from all sources of nitrogen pollution.

BNF is 'biological nitrogen fixation', a process through which bacteria associated with plant roots and in the soil convert di-nitrogen into reactive nitrogen compounds.

Überall und unsichtbar, lebensnotwendig und toxisch: Justus Liebig hatte in den 1840er Jahren die Idee und experimentierte damit, wie man Stickstoff aus der Luft nützlich und ertragssteigernd in den Boden bringen könnte, Fritz Haber und Carl Bosch gelang es zu Beginn des 20. Jahrhunderts, den Stickstoff in Ammoniak und damit in für Pflanzen verwertbaren Stickstoff umzuwandeln. Liebigs Antrieb war die Vermeidung von Hungersnöten in einer wachsenden Weltbevölkerung, die zunehmend in Städte lebte. Die entsprechende Vermehrung der Lebensmittel war sein Gegenmittel, während der auch kurz in Gießen weilende Georg Büchner im "Hessischen Landboten" den Palästen den Krieg erklärte und die Bauernrevolution auf die Tagesordnung setzte.[1] Gewonnen hat einstweilen die unter anderem mit dem Haber-Bosch-Verfahren eingeleitete "grüne Revolution". Noch vor wenigen Jahren glaubten die Vereinten Nationen, das Problem des Hungers endgültig von der Liste der größten Menschheitssorgen streichen zu können. Es kann deswegen nicht gestrichen werden, weil die Verarbeitung von Stickstoff und damit einhergehende Folgeprobleme massive ökologische Schäden hervorgerufen hat, die auch die Versorgung mit Lebensmitteln bedrohen. Die Menschheit ist auch in diesem Bereich des wissenschaftlich-technischen Fortschritts an planetare Grenzen gezogen und muss Bilanz ziehen.[2]

Segen und Fluch lagen schon im Ersten Weltkrieg zusammen, als aus Ammoniak auch Unmengen tödlichen Sprengstoffes erzeugt wurden. Ein Problem wurde dann vor allem die Aufblähung der Stickstoffmengen in den Kreisläufen, verschärft durch das NOx-Problem, die Verbrennung von fossilen und nachwachsenden Rohstoffen. Nitrat im Grundwasser und Lachgas in der Atmosphäre sind jedem Laien ein Begriff, die Verwendung von Kunstdünger ist drastisch einzuschränken. Und da sind wir fast wieder bei Liebigs Ausgangsproblem: wie eine immer noch wachsende Bevölkerung zu ernähren ist, angereichert durch die Existenz wichtige Frage, wie dies auf nachhaltige und gesunde Weise zu schaffen ist. Die technischen Kompensationsmöglichkeiten dazu gibt es, nationalen und europäischen Regulierungsinstanzen bestehen, quer stellen sich anachronistisch gewordene und durch starke, aber kurzsichtige Lobbygruppen am Leben erhaltene Strukturen und Organisationsformen der Agrarproduktion.

Das Stickstoff-Problem hat längst planetare Ausmaße angenommen und bedarf auch transnationaler, multilateraler Interventionen, es steht im Kreuzungspunkt so gut wie aller Umweltschäden (Wasser- und Luftqualität, Treibhausgase und Artensterben, Ozon) und betrifft fast alle Nachhaltigkeitsziele der Vereinten Nationen.

[1]Claus Leggewie, Justus Liebig trifft Georg Büchner: Sturm und Drang — eine fiktive Begegnung, in: Chemie in unserer Zeit, Bd. 37, H. 6, 2003, S. 410–415 https://doi.org/10.1002/ciuz.200300306.

[2]Rockström J, Steffen W, Noone K, Persson Å, Chapin F S, Lambin E F, Lenton T M, Scheffer M, Folke C, Schellnhuber H J, Nykvist B, De Wit C A, Hughes T, Van Der Leeuw S, Rodhe H, Sörlin S, Snyder P K, Costanza R, Svedin U, Falkenmark M, Karlberg L, Corell R W, Fabry V J, Hansen J, Walker B, Liverman D, Richardson K, Crutzen P, Foley J A (2009): Planetary Boundaries: Exploring the safe operating space for humanity, Ecology and Society 14 (2): http://www.stockholmresilience.org/planetary-boundaries.

Unter der Ägide von M. Sutton vom Centre for Ecology & Hydrology (CEH)[3] in Edinburgh ist 2019 im Auftrag des International Nitrogen Management System (INMS) noch einmal eine umfassende Problemanalyse vorgelegt worden, die in einen Alarmruf der Wissenschaftler'innen mündet:

The report is framed as a manifesto for "Science-in-Action". By this, we mean applying the evidence from science and the experience of researchers engaged in policy development to develop a vision of how Sustainable Nitrogen Management must be a core theme for sustainable development. The report is aimed at stimulating discussion, raising awareness and catalyzing action by governments, business, civil society, scientists and citizens for a more sustainable Nitrogen Earth. We present options as opportunities for environment, health and economy. In identifying these opportunities, we point to benefits and limitations of particular actions, while avoiding recommendations for specific policies. These must be for governments to agree, locally, regionally, nationally and internationally, in the light of the emerging evidence".

Möge dieses Buch dazu einen Beitrag leisten.

Claus Leggewie was born in 1950 in Wanne-Eickel. Ludwig Börne-Professor, University Gießen, Head of the Panel on Planetary Thinking. Until 2017 Director Institute of Cultural Sciences, Essen, Germany, and until 2016 member of the advisory council, German Federal Republic, Global Environmental Changes (WBGU).

Prof. Leggewie teached and researched in Paris-Nanterre, Algeria, at the Department of Human Science (IWM), in Vienna and New York University, where Prof. Leggewie received in 2016 the Sanders Prize. Prof. Leggewie is honorary doctor of the University Rostock and co-editor of the Journal Transit. He published Jetzt! Opposition, Protest, Widerstand, Cologne, 2019, and works presently on the subject "Politische Kunst im Anthropozän".

[3]Sutton M A, Howard C M, Adhya T K, Baker E, Baron J, Basir A, Brownlie W, Cordovil C, de Vries W, Eory V, Green R, Harmens H, Hicks K W, Jeffery R, Kanter D, Lassaletta L, Leip A, Masso C, Misselbrook T H, Nemitz E, Nissanka S P, Oenema O, Patra S, Pradhan M, Ometto J, Purvaja R, Raghuram N, Ramesh R, Read N, Reay D S, Rowe E, SanzCobena A, Sharma S, Sharp K R, Skiba U, Smith J U, van der Beck I, Vieno M, and van Grinsven H J M (2019): Nitrogen grasping the challenge. A manifesto for science-in-action through the International Nitrogen Management System, Summary Report. Centre for Ecology & Hydrology, Edinburgh, UK.

Editorial

Most people are completely unaware of the technical N_2 fixation (TNF) invention in the year 1909 by Haber and Bosch. TNF invention is prime example for one of the absolute foundation stones of modern existence and enables chemistry to develop N containing polymers which are worldwide used (Chap. 1) and agriculture keeping over half of the human race alive by overplaying the nitrogen shortage in food production.

Before TNF invention the N_2 fixing bacteria and archaea on Earth with all their growth limitations in waters and soils alone were qualified to keep large swathes of the human race alive. Such creatures developed long before Haber–Bosch's invention the idea to couple N_2 fixation with the photosynthesis process. The direct use of sun's energy to convert N_2 and H_2 into NH_3 is basis of enzymes, needed to form biomass. The Haber and Bosch invention, based on the pre-exercised nature's thoughts, allows synthesizing in huge reactors all over the world nearly 300 tons of ammonia per minute by using sun's energy stored in mineral oil. All over the world running NH_3 synthesizing reactors convert with mineral oil generated high temperatures, carefully-worked-out iron-based catalysts and under high pressures N_2 and H_2 into NH_3 and would the such driven reactors stop working the human population would looking at mass famines within months and would have a decent shot at collapsing our civilization, not seen since the Black Death.

NH_3, unimportant whether biologically or technically produced by N_2 fixation, channels through a complex of separate functioning bacteria, archaea, fungi, protozoa, nematodes, and all other animals and plants. NH_4^+ incorporated into proteins is oxidized to NO_3^- and NO_3^- after reduction to N_2 under energy gain returns to the atmosphere (Chaps. 1–5). On the cycling of fixed N_2 Nature's productivity is based. A high biodiversity can develop by having long term adapted on N shortage, because biological N_2 fixation has a limited NH_3 production capacity and curtails monoculturing agriculture in its productivity (Fig. 1).

TNF invention enabled monoculturing agriculture to surpass the NH_4^+ plant demand, driving the nitrifying bacteria, archaea consortium to top efficiency. The synthetic NH_4^+ chemistry overplays the soil biodiversity roles in agricultural fields by consuming high amounts of in oil stored energy (Chaps. 2–5). The wealth of

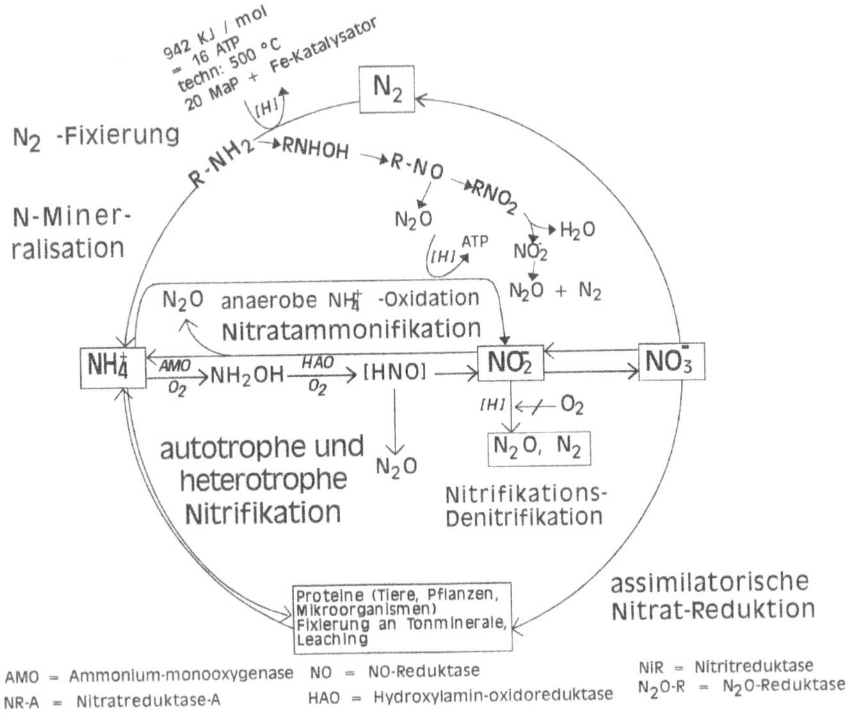

Fig. 1 The N cycle and its functioning, based on diverse groups of N_2 fixing, N immobilizing, nitrifying, and denitrifying bacteria and archaea (Benckiser, 1997)

mobile nitrate ions from which plants benefit enables the soil bacteria, archaea, fungi consortium to respire at developing anaerobic environments with NO_3^- by gaining energy and reducing NO_3^- over NO_2^-, NO and N_2O to N_2. Has the nitrate respiring soil consortium surplus NO_3^- available it starts stopping the reduction process at the N_2O level, despite N_2O has a highly positive redox potential (Chap. 2). NO_3^- reduction occurs less economically and less N_2O converts under energy gaining aspects and by accepting NH_3 derived electrons and protons to climate neutral N_2. World's atmosphere becomes susceptible to warm up. In this context of interest is that not only denitrifying bacteria and archaea possess the *nos*Z gene that is responsible for the reduction of N_2O to N_2 under energy gain (Chaps. 2 and 3; [4]).

Almost contemporaneous to the Haber–Bosch invention Heinrich Dreyer invented 1915 the drum caster (Fig. 2).

The patented TNF and drum-caster inventions were starting points for a steadily increasing TNF fertilizer spreading by reaching plant demand surpassing dimensions and over the past 100 years agriculturally used soils have received millions tonnes of N in form of urea, ammonium, nitrate (Fig. 3), forcing the long term on N shortage adapted soil biodiversity to change the living style [17].

Fig. 2 The drum caster for distributing nitrogen fertilizer, invented 1915 by Heinrich Dreyer

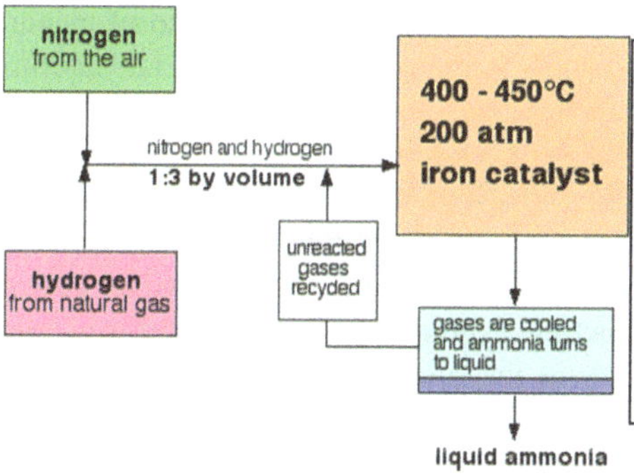

World demand for N fertilizer
And ammonia supply 2014-2019
(thousand tonnes)

Year	Nitrogen (N)	Ammonia-N
2015	112 539	153 766
2016	113 955	159 490
2017	115 498	164 724
2018	116 905	168 056
2019	118 222	171 433

Fig. 3 How the conversion of N_2 into NH_3 works and World' N fertilizer demand (Food and Agriculture Organisation, FAO, Summary report ISBN 978-92-5-108692-6 (http://www.fao.org/publications)

Asides TNF and on fields left organic residues agriculturally used soils must uptake increasing amounts of N, stored in manure from an intensified industrial livestock farming and in sewage of wastewater treatment plants (WWPTs). In addition a vast array of landscapes and aquatic systems N containing chemicals (nylon, polyurethanes, polyacrylonitriles, sulphonamide antibiotics, insecticides stress Chaps. 1–4) [3]. To all the N in manure, sewage and chemicals a flood of 8.3 billion metric tons of plastic materials, composed out of the elements carbon hydrogen, oxygen, nitrogen, sulphur chlorine (Chap. 1) release besides climatic events, plants and the above, belowground bacterial, archaeal, fungal, and faunal communities foot prints in soil profiles, visualizing and memorizing past and ongoing climate events, soil properties dependent biological activities and TNF production footprints.

Scientists, metabolic concept decoders, and synthetic biology designers try to improve the application of TNF and to get new ideas towards detecting, optimizing, and manipulating organisms as the nitrifying isolate Ca. *Nitrospira inopinata* [7, 15, 17]. Chapter 6 discusses how farmers could progress in employing precision or organic farming for achieving a more plant N demand adapted fertilization in monocultures, cropped to few, high-yielding varieties. In increasingly into mono-culturing converting landscapes pollinating insects find less food, bio-diversities are decreasing, and more ecologically based management strategies are claimed [9, 12]. The politic is forced to handle, the more because in cities concentrating and through TNF better nourished people contribute with their N loads to overused landscapes, oceans, and the atmosphere. Although a progressing WWPT technology could reduce the N input into Berlin's river Havel or into the central Germany crossing river Weser from 6500 to 3500 (Havel) and from 74,000 to 29,000 tons (Weser), German regions as the 320,000 hectares agricultural land in Lower Saxony, that the river Weser is crossing, in addition suffer under an industrialized pork and poultry production, *inter alia* fed with worldwide produced soya. Between 1700 and 1980 agricultural land expanded by 466% and the annual soil N surplus of 100 kg per hectare on the soil N buffer capacity drastically surpasses and disposed of manure amounts the above, belowground soil bacterial, archaeal, fungal, and faunal communities must adapt [7, 9, 12, 16]. In consequence the European Union policy suggests a N input reduction to 50 kg N per hectare that the ground water is less polluted with nitrate and the atmosphere with NO and N_2O [4]. On the question: can organic and precision farming concepts help correcting the accidentally by Haber–Bosch caused nowadays situation the book Chap. 6 tries to give an answer and recycling scientists started constructing synthetic genomes at the pro- and eukaryotic organism level to understand (a) the rules of life, (b) cell growth and programmed cell death, and (c) to progress in cultivating organisms by in situ observations with molecular biological techniques (Table 1).

In the pre-Haber–Bosch period organisms found ways to adapt on N shortage, but also learnt in the neighbourhood of cadavers *inter alia* to manage N surplus and r- and k-strategists developed (r = maximal intrinsic rate of natural increase; K refers to carrying capacity). Orientating on the by nature pre-lived geniality a new, better future, an improving the designing of proteins, enzymes production, and industry economy is tried (Chaps. 4–6; [5–6, 8, 13]).

Table 1 Cultivation progress in %

Habitate	(%)
Sea water	0.001–0.1
Mesotrophe lakes	0.1–1
Estuary	0.1–3
Sediments	0.25
Soil	0.3
Activated sludge	1–15
Human faeces	20–40
Oral cavity	>50

Amann et al. [1]

Fig. 4 Soil profile of a loess derived, weakly compacted, pseudogleyic Parabrown earth (Luvisols, Alfisol FAO, USDA Soil Taxonomy) with a grey-white compact, wet bleached, cations depleted, red- brown, clay–humus enriched, water retaining subsoil with grey sections under forest, Kinzig–Murg region, Germany, Forest Soils [11], published by the Forestry Administration of the German state Baden–Württemberg

References

1. Amann, R.I., Ludwig, W., Schleifer, K.H.: Phylogenetic identification and in situ detection of individual microbial cells without cultivation. Microbiol. Rev. **59**(1), 143–169 (1995)
2. Benckiser, G.: Plastics, micro- and nanomaterials, and virus-soil microbe-plant interactions in the environment. In: Prasad, R. (ed.) Plant Nanobionic, vol. 1, pp. 83–96. Springer Nature Switzerland (2019) (Chapter 4)
3. Benckiser G.: Biofilm surrounded hexachlorobenzene (HCB) crystals and wastewater purification. Examines Mar Biol Oceanogr. **3**(5). EIMBO.000574 (2020) https://doi.org/10.31031/EIMBO.2020.03.000574
4. Benckiser, G., Ladha, J.K., Wiesler, F.: Climate change and nitrogen turnover in soils and aquatic environments. In: J. Marxsen (ed.) Climate Change and Microbial Ecology: Current Research and Future Trends, p. 22. Caister Academic Press, Norfolk, UK (2016) (Chapter 8)
5. Berry, D., Mader, E., Lee, T.K., Woebken, D., Wang, Y., Zhu, D., Palatinszky, M., Schintlmeister, A., Schmid, M.C., Hanson, B.T., Shterzer, N., Mizrahi, I., Rauch, I., Decker, T., Bocklitz, T., Popp, J., Gibson, C.M., Fowler, P.W., Huang, W.E., Wagner, M.: Tracking heavy water (D_2O) incorporation for identifying and sorting active microbial cells. PNAS **112**, E194–E203 (2015). https://doi.org/10.1073/pnas.1420406112Photomicrographs
6. Blount, B., Ellis, T.: Genome construction amends building codes. Nature **569**, 492–494 (2019). https://doi.org/10.1038/d41586-019-01584-x
7. Chen, J.G., Crooks, R.M., Seefeldt, L.C., Bren, K.L., Bullock, R.M., Darensbourg, Y., Holland, P.L., Hoffman, B., Janik, M.J., Jones, A.K., Kanatzidis, M.G., King, P., Lancaster, M., Lymar, S.V., Pfromm, P., Schneide, W.F., Schrock, R.R.: Beyond fossil fuel–driven nitrogen transformations. Science **360**, 6391 (2018). https://doi.org/10.1126/science.aar6611
8. Daims, H., Lebedeva, E.V., Pjevac, P., Han, P., Herbold, C., Albertsen, M., Jehmlich, N., Palatinszky, M., Vierheilig, J., Bulaev, A., Kirkegaard, R.H., von Bergen, M., Rattei, T., Bendinger, B., Nielsen, P.H., Wagner, M.: Complete nitrification by Nitrospira bacteria. Nature **528**(7583), 504–509 (2015). https://doi.org/10.1038/nature16461. Epub 26 Nov 2015
9. Eurostat Regional Yearbook Edition (2018). https://doi.org/10.2785/220518
10. Geyer, R., Jambeck, J.R., Law, K.L.: Production, use, and fate of all plastics ever made. Sci. Adv. **3**(e1700782), 1–5 (2017)
11. Glatzel, K., Jahn, R., Müller, S., Schlenker, G., Werner, J.: Südwestdeutsche Waldböden im Farbbild Schriftenreihe der Landesforstverwaltung Baden Württemberg, Germany, Band 23 (1967)
12. Habel, J.C., Ulrich, W., Biburger, N., Seibold, S., Schmitt, T.: Agricultural intensification drives butterfly decline. Insect Conserv. Div. (2019). https://doi.org/10.1111/icad.12343
13. Hahn, K.M.: An 'on' switch for proteins. Nature **569**, 490–491 (2019). https://doi.org/10.1038/d41586-019-01394-1
14. Jambeck, J.R., Geyer, R., Wilcox, C., Siegler, T.R., Perryman, M., Andrady, A., Narayan, R., Law, K.L.: Plastic waste inputs from land into the ocean. Science **6223**(347), 768–771 (2015)
15. Kitzinger, K., Padilla, C.C., Marchant, H.K., Hach, P.F., Herbold, C.W., Kidane, A.T., Könneke, M., Littmann, S., Mooshammer, M., Niggemann, J., Petrov, S., Richter, A., Stewart, F.J., Wagner, M., Kuypers, M.M.M., Bristow, L.A.: Cyanate and urea are substrates for nitrification by Thaumarchaeota in the marine environment. Nat. Microbiol. **4**, 234–243 (2019). https://doi.org/10.1038/s41564-018-0316-2
16. Matson, P.A., Parton, W., Power, A.G., Swift, J.: Agricultural intensification and ecosystem properties. Science **277**, 504–509 (1997). https://doi.org/10.1126/science.277.5325.504z
17. Sutton, M.A., Howard, C.M.: Ammonia maps make history. Nature **564**, 49–50 (2018). https://doi.org/10.1038/d41586-018-07584-7
18. Triplett, E.W. (ed.): Prokaryotic nitrogen fixation. Horizon scientific Press, Wymondham, UK (2000)
19. Van Damme, M., Clarisse, L., Whitburn, S., Hadji-Lazaro, J., Hurtmans, D., Clerbaux, C., Coheur, P.F.: Industrial and agricultural ammonia point sources exposed. Nature **564**, 99–103 (2018). https://doi.org/10.1038/s41586-018-0747-1

Contents

Recent Advances in Understanding the Role of Wastewater Treatment Processes for the Removal of Plastic Derived Nitrogen Compounds in Municipal Landfill Leachate

Kandasamy Ramani, Maseed Uddin, Krishnan Venkatesan Swathi, Rajasekaran Muneeswari, and Mohan Thanmaya

Abstract This chapter discusses the various wastewater treatment processes for the removal of nitrogen emissions from the plastic compounds present in municipal landfill leachate. The chapter briefs on the types of plastics and the means of plastic degradation that result in the emission of nitrogen into the waste stream and also describe the major state of the art of some of the most common biological processes such as Partial nitrification process, Anaerobic ammonia oxidation (ANAMMOX), Completely autotrophic nitrogen removal over nitrite (CANON), NO_x process and Oxygen-limited nitrification and denitrification (OLAND) process were discussed. In addition, conventional physicochemical methods such as break-point chlorination, ion exchange, membrane processes, precipitation and stripping are also summarized. Also, this chapter reviewed on the future challenges and perceptions on the mitigation of the nitrogen contamination by using advanced sustainable and eco-friendly plastic resources.

Keywords Municipal landfill leachate · Types of plastics · Plastics degradation · Plastic derived nitrogen compounds · Nitrogen emission · Biological processes

1 Introduction

Plastics transformed medicine with life-saving equipment, lightened cars and jets, made space exploration a reality and saved millions of lives with helmets, incubators, and devices for clean drinking water. The amenities plastics paved the way to a throw-away mentality that exposes the dark side of this material. Today, single-use plastics account for 40% of the plastic produced every year [1]. Plastic pollution is one of the most persistent environmental problems, as swiftly growing manufacture

K. Ramani (✉) · M. Uddin · K. V. Swathi · R. Muneeswari · M. Thanmaya
Biomolecules and Biocatalysis Laboratory, Department of Biotechnology, School of Bioengineering, SRM Institute of Science and Technology, Kattankulathur-603203, Chengalpattu District, Tamil Nadu, India
e-mail: ramanik@srmist.edu.in

G. Benckiser (ed.), *Soil and Recycling Management in the Anthropocene Era*, Environmental Science and Engineering,
https://doi.org/10.1007/978-3-030-51886-8_1

of disposable plastic products overpowers the world's capacity to deal with them. Developing Asian and African nations are where plastic pollution is most discernable, mainly due to inefficient or absent garbage collection systems. But developed countries are also facing trouble in properly collecting discarded plastics due to low recycling rates. Plastic trash has become so pervasive that it has driven efforts to write a global treaty negotiated by the United Nations. American Association for the Advancement of Science (AAAS), has come out with a new observation on plastic, which points out that 79% of the total plastic waste of 6300 million metric tons (MMT) is accumulated in landfills or in the natural environment like river systems and oceans. The study, 'Production, Use and Fate of all Plastics Ever Made', highlighted that if existing creation and waste management tendencies endure then around 12,000 MMT of plastic waste will end up in landfills or in the natural environment by 2050. There are evolving technologies like pyrolysis, which extracts fuel from plastic waste [2]. But mostly all thermal degradation has been by incineration, with or without energy recovery. The environmental and health impacts of waste incinerators sturdily hinge on the emission control technology, incinerator strategy and maneuver [3]. Plastics can be castoff and either contained in a managed system, such as landfills, or left uncontained in open junkyards or in the natural environment.

The conventional (fossil fuel-based) plastic waste is non-biodegradable and remains in the soil for several years causing immense pollution in the environment as life cycle of plastic waste is partial and eventually it is dumped on the land-fill locations. It is well recognized that all kinds of plastic wastes cannot be recycled. Therefore, it gets concentrated in open drains, low-lying areas, river banks, coastal areas and sea-beaches which would inevitably affect soil, ground water and the surroundings [4]. Here the plastics under environmental stresses degrade into smaller plastic particulates thereby releasing a number of toxic substances in the process. It is projected that there are already 165 million tons of plastic debris moving around freely in the oceans which are a major threat to the health and safety of marine life. An average of 8.8 million more tons of plastics enter the oceans each year which includes microplastics, tiny particles less than five millimeters long from cosmetics, fabrics or the breakdown of larger pieces, which may be ingested by marine wildlife [5]. Most of the plastic trash end up in the oceans because they are carried to sea by major rivers that flows through land which act as conveyor belts, picking up more and more trash as they move downstream. Among the plastics used, nitrogen-containing plastics are extensively used for various products and constitute approximately 10% of total plastic usage [6]. The breakdown of plastics causes the desorption and release of various chemical compounds which are used as additives in the production of plastics to increase its durability and flexibility. Various volatile and semi volatile components are released during the degradation of nitrogen containing plastics [7]. Some of these compounds are nitrogenous in nature. They pose great environmental risks and proper and efficient wastewater treatments should be used to counter this menace.

1.1 Types of Plastics

1.1.1 Polyamide

Polyamides are made of amide monomers connected together with peptide bonds. Polyamides can be either aliphatic or aromatic. Polyamide 6 contains around 11.2% of nitrogen. Lauric acid, Myristic acid, Palmitic acid and Stearic acid are some of the chemicals that are released on the degradation of polyamide 6.

1.1.2 Polyurethane

Polyurethane (PUR and PU) is a polymer composed of organic units joined by carbamate (urethane) links. Some of the compounds released during the degradation of polyurethane are Lauric acid, Myristic acid, Palmitic acid, Diethyl phthalate, Dibutyl phthalate and Di-isobutyl phthalate.

1.1.3 Urea Formaldehyde

Urea-formaldehyde (UF), also known as urea-methanal, is a non-transparent thermosetting resin or polymer. This polymer is made from a combination of urea and formaldehyde. Chemicals like Lauric acid, Myristic acid, Palmitic acid, Methyl palmitate, Stearic acid and Methyl stearate are released when urea formaldehyde breaks down.

1.1.4 Melamine Formaldehyde

Melamine resin or *melamine formaldehyde* is a *resin* with *melamine* rings terminated with multiple hydroxyl groups which are derived from *formaldehyde*. This thermosetting plastic material is made from *melamine* and *formaldehyde*. Lauric acid, Myristic acid, Palmitic acid, Methyl palmitate, Stearic acid, Methyl stearate, Diethyl phthalate, Dibutyl phthalate, Diethylhexylphthalate and Butylated hydroxytoluene are some of the compounds that are desorbed on the degradation of melamine formaldehyde.

1.1.5 Acrylonitrile–Butadiene–Styrene

Acrylonitrile-butadiene styrene (ABS) is a common thermoplastic polymer. ABS is a terpolymer made by polymerizing styrene and acrylonitrile in the presence of polybutadiene. Compounds like Diethyl phthalate, Dibutyl phthalate, Diisobutyl phthalate, Diethylhexylphthalate and Butylated hydroxytoluene are released when ABS is degraded.

1.1.6 Polyethylene, Polypropylene and Polystyrene

Polyethylene, Polypropylene and Polystyrene are types of plastics that do not contain nitrogen in its structure but still releases nitrogen containing compounds during its degradation namely drometrizole and benzothiazole, which possibly are included as additives for ultraviolet stabilization and vulcanization acceleration, respectively [6].

1.2 Methods of Plastic Degradation

Polymer degradation involves bond scissions and subsequent chemical transformations that are due to the chemical, physical or biological reactions causing changes in polymer properties. Degradation mirrors changes in material properties such as mechanical, optical or electrical characteristics. Polymer degradations have been classified as photo-oxidative degradation, thermal degradation, ozone-induced degradation, mechanochemical degradation, catalytic degradation and biodegradation depending upon the nature of the causing agents.

1.2.1 Photo-Oxidative Degradation

The process of decomposition of a material by the action of light is called Photo-oxidative degradation. Even at optimal conditions photo oxidation is considered as one of the chief causes of damage exercised on polymeric substrates. Most of the synthetic polymers are prone to breakdown initiated by UV and visible light. Photochemicalreactions occur only on the surface of the polymer. The physical composition of polystyrene can be greatly decreased through widespread cleaving of the polymer chains during photo degradation [8].

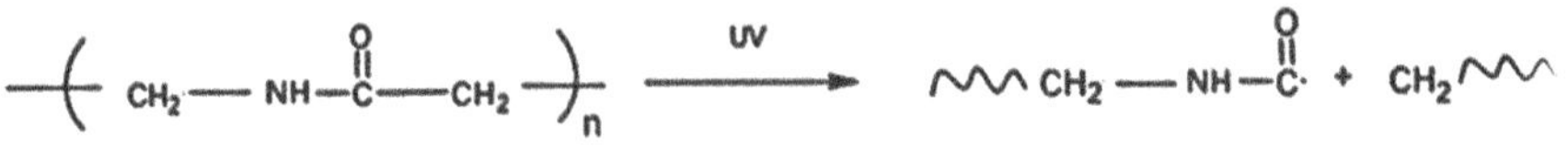

1.2.2 Thermal Degradation

Under ordinary environments, photochemical and thermal degradations are similar and are classified as oxidative degradation. Thermal reactions occur throughout the bulk of the polymer sample [9]. Random and chain degradation introduced by thermal and UV light cause thermal degradation of the polymers.

1.2.3 Ozone-Induced Degradation

Normal conditions, degradation of polymers can occur because of atmospheric ozone as well. This method can be adopted when other oxidative aging processes are relatively slow and the polymer holds its properties for a rather longer time. The presence of atmospheric ozone hastens the aging of polymeric materials and it is convoyed by the rigorous formation of oxygen-containing compounds.

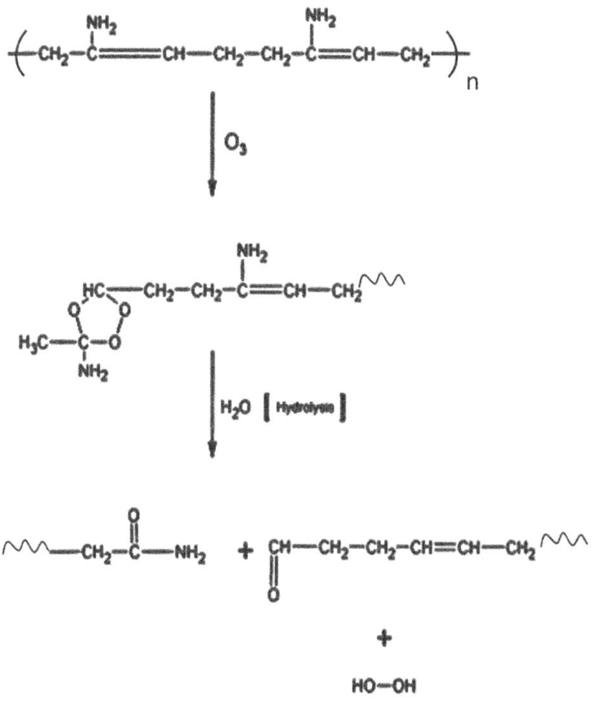

1.2.4 Mechanochemical Degradation

The degradation of polymers under mechanical stress and by strong ultrasonic irradiations comes under mechanochemical degradation. Breakdown of molecular chains under shear or mechanical force is often accompanied by a chemical reaction.

1.2.5 Catalytic Degradation

Catalytic transformation of polymeric substances into hydrocarbons with higher commercial value can be greatly advantageous. Gases and oils can be extracted by thermally or catalytically degrading polyolefins. A lot of attention has been directed towards polyolefins (PE, PP, PS) which comprise an important part of industrial and domestic waste. The addition of a catalyst not only improves the quality of products obtained from pyrolysis of plastic wastes and lower the temperature of decomposition, but also enables high selectivity for the products to be achieved. Free radical mechanism can be adopted for the catalytic degradation of the polypropylene using Fe activated carbon catalyst [10].

1.3 Biodegradation

Biodegradation comprises of the biochemical breakdown of compounds by microorganisms. Under aerobic conditions mineralization of organic compounds yields carbon dioxide and water whereas methane and carbon dioxide are released under anaerobic conditions. Biodegradation of polymers can be enhanced by abiotic hydrolysis, photo-oxidation and physical disintegration which in turn would increase their surface area for microbial colonization [11]. Biodegradation causes a change in surface properties leading to a loss of mechanical strength, degradation by enzymes and the final decrease in the average molecular weight of the polymers. Degradation can occur as a combination of these mechanisms or even individually. Depending on the mechanism biodegradation can occur at microscopic, macroscopic, molecular and macromolecular levels.

1.3.1 Solubilization

The hydrophilicity of the polymer plays a huge role in the hydration of polymers. The secondary and tertiary structures of the polymers stabilized by van der Waals forces and hydrogen bonds are disturbed on hydration. For polymers having crosslinks the breakage of either the polymer backbone or crosslinker can reduce its strength [12]. A bio-surfactant that increases the solubility of polyaromatic hydrocarbons can also be used in the biodegradation of polymers.

1.3.2 Ionization

Ionization or protonation of a pendent group can help in solubilizing certain water insoluble polymers. Polyacids can be made soluble and hence hydrophilic.

1.3.3 Hydrolysis and Enzyme-Catalyzed Hydrolysis

Polymers that are insoluble in water and containing ester groups can be made soluble if the esters are hydrolyzed to form ionized acids on the polymer chain [4]. Enzymes can also be used as catalysts for oxidation, reduction, hydrolysis and esterification [13].

1.3.4 Microbial Degradations

One of the approaches to enhance biodegradability is to add natural polymers to thermoplastics. These natural polymers could be made from bacteria, fungi, algae, etc. Preparing biodegradable plastic encompasses addition of distinct additives to the synthetic polymers, which make it prone to microbial degradation by breaking the endurance of C–C chain of polymers [14].

1.4 Factors Affecting Polymer Degradation

1.4.1 Chemical Composition

The chemical configuration of the polymers is of great importance in its degradation. Breaking down of polymers by microorganisms can be hindered by the presence of polymers that contains only long carbon chains. Polymers can be made more susceptible to biodegradation by introducing hetero-groups such as oxygen in the polymer chain [15].

1.4.2 Molecular Weight

The rate of plastic degradation reduces with increase in the molecular weight of the plastics [16]. Linear plastic polymers with molecular weight lower than 620 can support microbial growth [17].

1.4.3 Size of the Molecules

Biodegradation of polymers are heavily influenced by the size of the molecules. The degradation rate increases as the size of the molecule reduces.

1.4.4 Functionality

The effect of hydrophilic moieties in the degradation of the polymers was studied by Singh and Sharma. Biodegradation was observed in pure PS and grafted PS and the findings were compared. It was found that a degradation of 37% was achieved in PS when the polymer was adapted with starch after 160 days and no degradation was seen in grafted PS with poly (AAc) [4]. This was primarily because starch is a natural hydrophilic polymer and hence is more vulnerable for bacterial and fungal degradations in soil if optimum environment is presented for their growth.

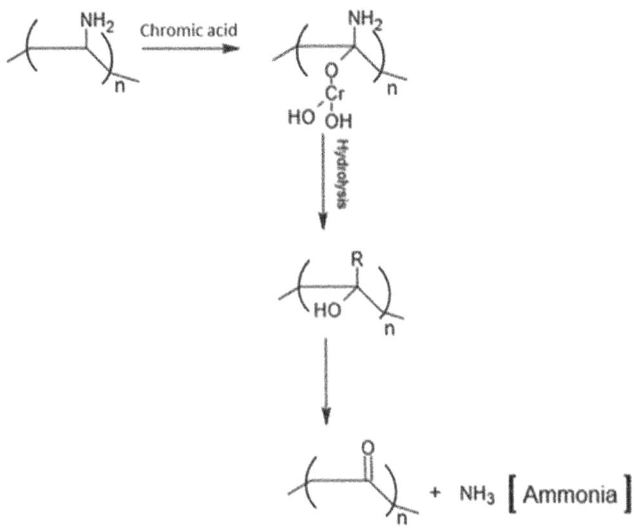

1.4.5 Additives

Fillers, pigments and non-polymeric impurities like polymerization catalysts residues, affect breakdown of plastics by causing undue resistance to degradation [18]. But some metals can act as a good pro-oxidant in polyolefins making the polymer more liable to thermo-oxidative degradation.

1.4.6 Chemical Bonding

The chemical linkages and bonding in polymers can influence the rate of degradation in plastics. Head-to-head and tail-to-tail addition of monomer units during addition polymerization can form weak joints which make thermoplastics more prone to degradation [19]. Thermal degradation and photodegradation increases with branching in the polymer chain and decrease with crosslinking as it averts lamellar unfolding.

1.4.7 Effect of Stress

Stress can have a major influence on polymer disintegration. The rate of photodegradation increases with increase in tensile stress and decreases with compressive stress. The morphology of the plastic polymer can be studied to infer the consequence of stress on degradation rates. The straightening of the polymer chains in the amorphous regions can be observed when high stress is being subjected on the polymers. Stressed polymer chains had an increased rate of crack development and hence underwent a higher degree of degradation as compared to unstressed state [20].

1.4.8 Environmental Conditions

Biodegradation of polymer depends upon environmental settings like pH, temperature, amount of moisture, oxygen content, and required population of microorganisms [21]. The rate of polymer degradation by the microorganisms increases with increase in relative humidity ideally above 70%. The combined effect of temperature and moisture demonstrates a significant degree of photodegradation of the polymeric materials. Higher temperatures and high humidity have been studied to cause an increase in the photo-damage in thermoplastic polymers [22].

2 Biological Treatment of Wastewater Containing Plastic Derived Nitrogen Compounds

2.1 Heterotrophic Nitrification

It involves the use of chemoorganotrophic microorganisms which oxidize ammonia [23], organic nitrogen compounds [24] and hydroxylamine [25] to nitrite and nitrate. Commonly used heterotrophic nitrifiers are bacteria [26], algae [27] and fungi [28]. Nitrification rate of heterotrophic nitrifiers are low when compared to autotrophic nitrifiers [29], therefore heterotrophic nitrification is preferentially used under environment which is unfavourable for autotrophic nitrification, e.g. acidic environment.

2.2 *Denitrification*

It involves the use of chemoorganotrophic, phototrophic and lithoautotrophic bacteria and fungi which reduces the nitrite or nitrate to gaseous nitrogen compounds [30, 31] under anoxic or oxygen-reduced environment [32]. Denitrification is a kind of anoxic respiration, where electrons generated from molecular hydrogen, reduced sulphur compounds, or organic compounds are transferred to nitrite or nitrate instead of oxygen, which helps in ATP generation. Enzymes involved during the denitrification are the nitrite reductase, nitrate reductase, nitric oxide reductase and nitrous oxide reductase [33, 34]. Dinitrogen is the major end product of this process while nitrous oxide and nitric oxide are also produced in low concentrations. However, nitrous oxide and nitric oxide are also released as the end product of denitrification process when the concentration of dissolved oxygen is too high [35]. The onset of aerobic denitrification is dependent on the regulation of the redox-sensing factors which acts as a transcription regulator.

2.3 *Nitrogen Removal*

Nitrogen removal form the wastewater can also be accomplished by the use of newly discovered anaerobic metabolism of proteobacterial ammonia oxidizers and anaerobic ammonia oxidizing planctomycetes. In this section the application of the newly discovered microorganisms is discussed.

2.3.1 Partial Nitrification

It involves the oxidation of ammonium present in the wastewater to nitrite, not to nitrate and therefore subsequent oxidation of nitrite to nitrate must be prevented. In order to increase nitrogen removal efficiency, partial nitrogen removal can be combined with anammox process and also with conventional denitrification process. When combined with conventional denitrification process it gives significant advantage of resource utilization [36]. It requires less aeration since the subsequent denitrification step reduces nitrite to molecular nitrogen not the nitrate.

Oxidation of nitrite to nitrate is prevented by two different methods. First, by using difference in activation energy between nitrite oxidation and ammonia (44 kJ/mol and 68 kJ/mol respectively). The SHARON (Single reactor for high activity ammonium removal over nitrite) process use different growth rate of nitrite and ammonia oxidizers at high temperature (more than 26 °C) [37, 36] and runs at a hydraulic retention time higher than nitrite oxidizers growth rate but lower than ammonia oxidizers. SHARON process has no sludge retention therefore the nitrite oxidizers are washed out from this process. SHARON process is not suitable for all the wastewater since it depends on high temperature. Rotterdam wastewater treatment plant is a typical example of scaled up SHARON process for treatment of sludge liquor [38].

Second method which prevents the oxidation of nitrite to nitrate is the modified SHARON process which makes use of sludge retention [39]. It works at surplus ammonium and low oxygen concentration (less than 5% air saturation), which prevent the growth of nitrite oxidizers and thereby prevent the oxidation of nitrite and makes it stable. Mechanistic evidence for the impaired growth of nitrite oxidizers by ammonia and low oxygen concentration is still not clear.

2.3.2 Anaerobic Ammonia Oxidation (ANAMMOX)

It is a denitrification process of nitrite, employing the use of ammonia as electron donor [40, 41]. The process requires a nitrification step which converts ammonia to nitrite. Modified SHARON process has been used to produce ammonium/nitrite mixtures by removing the anoxic step and not supplying methanol [41, 42]. The first scaled up anammox reactor is installed in Rotterdam, Netherland. It is in addition to the SHARON process. Depending on the reactor design and the ammonium concentration of the wastewater, the dinitrogen gas produced during the process can be used to partially mix the reactor and thereby reducing the power consumption. The recycling part of the nitrogen gas can be used for additional mixing of the reactor. The reactor should be well mixed to prevent the formation of toxic sulphide and to maintain redox potential in denitrification zone. Reactor should not be overloaded because high nitrite concentration is detrimental for the microorganisms (more than 180 mg N/L NO_2^- for *Candidatus* Kueneniastuttgartiensis and more than 70 mg N/L NO_2^- for *Candidatus* Brocadiaanammoxidans) [43, 44]. Anammox process has been conducted on laboratory scale in different reactor: fluidized bed [44], fixed bed [39] and sequencing batch [45] and found to be suitable for all the processes. The major challenges of the anammox process is its long start up time, because of the slow growth rate of the anammox planctomycetes (100–150 days before the inoculation of the reactor with activated sludge) [42]. This problem may be overcome and seeding can become possible once the anammox plants are in full scale operation.

2.3.3 Canon

It stands for completely autotrophic nitrogen removal over nitrite. It is a combination of partial nitrification and anammox process in a single aerated reactor [46, 39, 47]. The word canon refers to cooperation of the two groups of bacteria: perform sequential reactions simultaneously (Eqs. 1 and 2). The nitrifiers group of organisms involves in the oxidation of ammonia to nitrite and consume oxygen and thereby create anoxic environment required for the anammox process. Canon process has been performed on laboratory scale. The volumetric loading rate of the canon process (1.5 kg N m^{-3} day^{-1} in a gas—lift reactor) is lower than anammox process [47]. However, the economics of the canon process might be advantageous because of the use of only one reactor. Canon process has not been tested on full scale, but it is known to occur in nitrification system and convert ammonium to nitrate and dinitrogen gas [48–50].

$$NH_4{}^+ + 1.5O_2 \rightarrow NO_2{}^- + 2H^+ + H_2O\left[\Delta G^{\circ\prime} - 275\,KJ\ mol^{-1}\right] \quad (1)$$

$$NH_4{}^+ + NO_2{}^- \rightarrow N_2 + 2H_2O\left[\Delta G^{\circ\prime} - 357\,KJ\,mol^{-1}\right] \quad (2)$$

2.3.4 NO_x Process

The addition of nitrogen oxides to the wastewater offers the new opportunity in the wastewater treatment by stimulating and controlling the denitrification activity of the nitrosomonas-like microorganisms. Addition of nitrogen oxide simulate the nitrosomonas-like microorganisms to nitrify and denitrify simultaneously under oxic condition with dinitrogen as the main product (Fig. 1). Equations (1)–(3) shows the nitrogen conversion in a subsequent nitrification and denitrification process without NO_x supply and Eqs. (6)–(8) shows the conversion of nitrogen which is influenced by NO_x supply. The [H] indicate reducing equivalents.

NO_x (NO/NO_2) acts as an inducers by inducing denitrification activity of the ammonia oxidizers. It is added in a trace amounts to the wastewater ($NH_4{}^+/NO_2$ ratio about 1000/1–5000/1) [51]. As a result 50% of the [H] (reducing equivalents) are transferred to nitrite (terminal electron acceptor) (Eq. 6) instead of oxygen, which results in the reduction in the oxygen consumption of the process. (Eqs. 6 and 8).

Nitrification

$$3NH_4{}^+ + 6O_2 \rightarrow 3NO_3{}^- + 6H^+ + 3H_2O \quad (3)$$

Denitrification

$$3NO_3{}^- + 3H^+ + 15[H] \rightarrow 1.5N_2 + 9H_2O \quad (4)$$

Sum

$$3NH_4{}^+ + 6O_2 + 15[H] \rightarrow 1.5N_2 + 12H_2O + 3H^+ \quad (5)$$

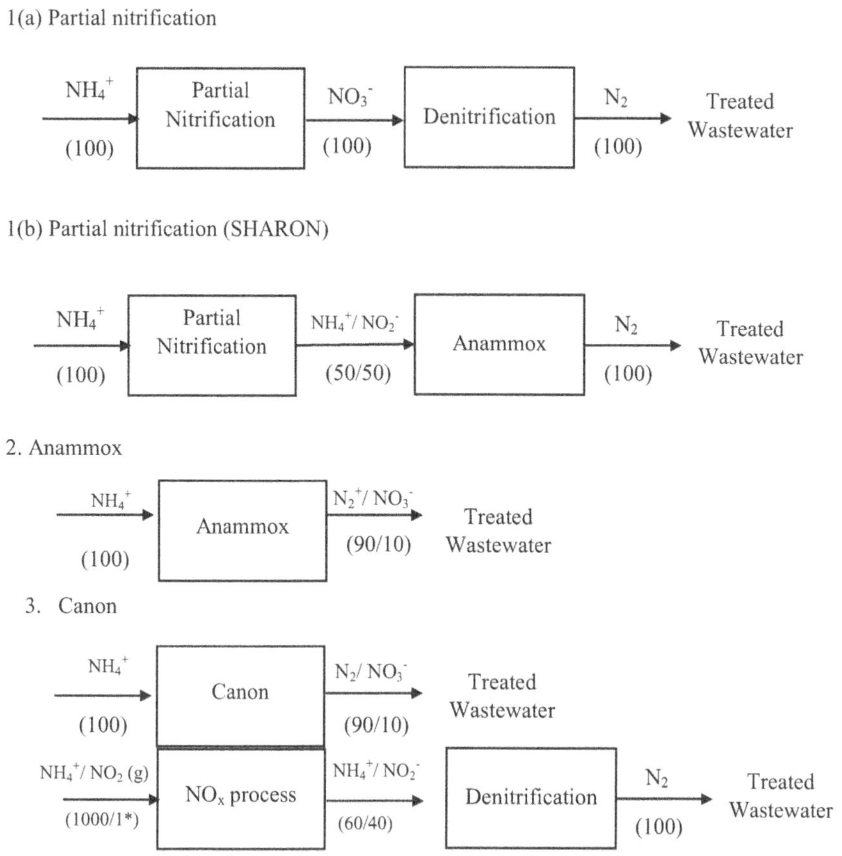

Fig. 1 Flux diagram of (1a) partial nitrification, (1b) SHARON, (2) Anammox, (3) Canon and (4) NO_x process

Plant with NO_x supply:

Nitrification

$$3NH_4^+ + 3O_2 \rightarrow N_2 + 4H_2O + NO_2^- + 4H^+ \tag{6}$$

Denitrification

$$NO_2^- + H^+ + 3[H] \rightarrow 0.5N_2 + 2H_2O \tag{7}$$

Sum

$$3NH_4^+ + 3O_2 + 3[H] \rightarrow 1.5N_2 + 6H_2O + 3H^+ \tag{8}$$

2.3.5 OLAND Process

Oxygen-limited nitrification and denitrification (OLAND) is a single step process involving the use of nitrifiers for the removal of ammonium without the addition of COD [52]. The ammonium loading rate of the process is low and the mechanism is yet to be explored. The OLAND process possibly be based on either Canon process (combine aerobic-anaerobic ammonia oxidizers) or NO_x process (nitrifier denitrification in the presence of NO_x). Aerobic deammonification is another single step process for the removal of ammonium without the addition of COD. After the complete study on pilot and full scale plant, it has been found that is based on the Canon process involving the use of nitrifiers and anaerobic ammonia oxidizers, working under oxygen limitation and thus converting ammonium to dinitrogen gas and nitrate (Fig. 1).

3 Physico-chemical Treatment of Wastewater Containing Plastic Derived Nitrogen Compounds

The weathering of nitrogen containing plastics in any environment releases nitrogen compounds into either ammonia nitrogen or nitrite and nitrate nitrogen. At present, a wide spectrum of physical and chemical processes is being followed for the elimination of nitrogen from the wastewater. For any particular application, the most suitable approach will be determined based on (i) the amount of nitrogen present in the wastewater; (ii) desired quality of effluent; (iii) other unit operations to be applied for the treatment of other pollutants present in the wastewater; (iv) economy of the process. In this section, some of the common physicochemical methods such as break-point chlorination, ion exchange, membrane processes, precipitation and stripping are individually discussed.

3.1 Breakpoint Chlorination

Breakpoint chlorination is a general procedure applied to recover nitrogen from the nitrogen containing effluent by oxidizing ammoniacal nitrogen by chlorination. Even though it is used widely in many N-ammonia wastewater treatments, the high operational cost and the process complexity remains a challenge. The advantages of this method are: (1) it is promising method to remove ammoniacal nitrogen present in the effluent to zero with application of adequate amount of (2) The low capital costs makes it particularly suitable for certain applications and it can be easily added to an existing treatment facility, where nitrogen removal is required [53].

During the treatment, the reaction starts with hypochlorous acid to yield monochloramines. As the addition of HOCL continues, the formation of nitrogen gas from the chloramines will occur once the free chlorine is liberated i.e., the total reaction is written below:

$$NH_4 + HOCl \rightarrow N_2 + H_2O + H^+ + Cl^-$$

Figure 2 illustrates the theoretical breakpoint curve occurred during chlorination process. In the Zone 1 region, the chloramine is formed when the added chlorine reacts with the ammoniacal nitrogen present in the effluent; Zone 2 relates an elevation in the concentration of dichloramine and the reduction in ammonia; Zone 3 shows liberation of chlorine next to the breakpoint. At Cl_2:NH_3 ratio equals to 5, drop in the ammonia concentration starts and it becomes nearly absolutely null at the breakpoint. Overall, it is established that monochloramine is being formed with an increase in the chlorine from zero to the "hump" in first zone, dichloramine is formed at the falling line of the breakpoint curve, and reduction in the NH_3–N shows the formation of N_2 gas [54] (Fig. 2).

Generally, 9–10 mg/l of chlorine is added to the effluent for the recovery of 1 mg/l of ammonia–Nitrogen and further the neutralization of acid products formed during the process is necessary. The application of the vast amount of chemicals in the process results in the building up of the total dissolved solids and substantially, increases the operational costs.

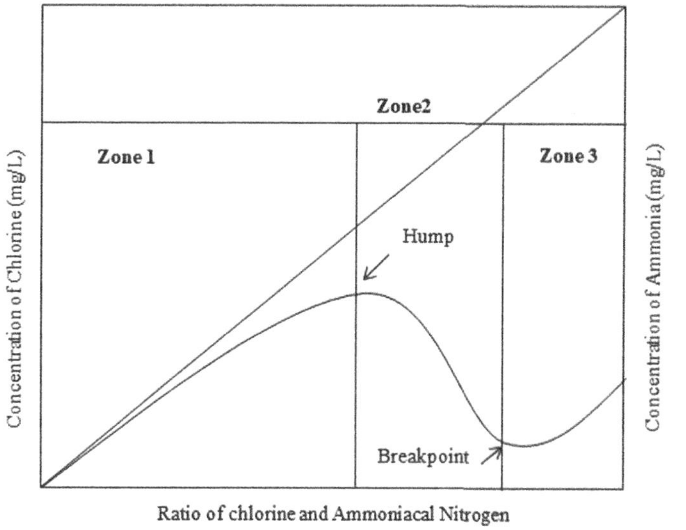

Fig. 2 Theoretical breakpoint curve

3.2 Sequential and Continuous Ion-Exchange Operation

In ion exchange operations, ions on the surface of a solid are exchanged for ions of an identical charge matrices present in the wastewater in which the solid is in contact. Ion exchange is a general process which can be used to remove undesirable ions from waste water. Cations (positive ions) are exchanged for hydrogen or sodium, and anions (negative ions) for hydroxide or chloride ions. Ion exchange using zeolites is a basic process applied for recovery of ammonia. Zeolites have a porous structure which is readily available for the trapping of the various cations. Many natural zeolites materials varying in Si and Al composition are available in which Si rich Clinoptilolite is the widely used ion exchange column. It has a dual channel system where the zeolite functions as a molecular sieve. The salient features of Clinoptilolite are high sorption and ion-exchange capacity, ion exchange selectivity, catalytic activity and structural temperature stability up to 700–750 °C. In addition, synthetic zeolites with some improved properties are also produced using silica and alumina as the main raw materials which further reduces the process economy [55].

The clinoptilolite based ammoniacal nitrogen removal is a cationic ion exchange reaction. In addition to the ion exchange, ammonia can also be recovered through the physical adsorption by the structural pores of zeolite. The temperature plays a main role in determining the efficiency of ammonia removal by adsorption in which highest ammonia removal is achieved at elevated temperatures. However, the maximum ammonia removal by clinoptilolite will occur between the pH 4 and 8. The presence of other cations such as potassium, ammonia, calcium, sodium, magnesium and other ions in the wastewater can easily be adsorbed by Clinoptilolite and other zeolite which further results in the diminishing ammonia removal ability [56] (Fig. 3).

3.3 Membrane Processes-Reverse Osmosis and Ultrafiltration

Membrane processes like electrodialysis, ultrafiltration and reverse osmosis play a significant role in the treatment of nitrogen containing waste water. A membrane is a phase that acts as a mechanical barrier between the other phases which can be of solid or a liquid. These efficiently used to the waste water containing nitrogen in the form of ammonium or nitrate.

Reverse osmosis is highly preferred among the other membrane processes since its maximum potential in high nitrogen removal and also capable of removing all forms of nitrogen. Approximately 60–90% of the total nitrogen is being removed by this process. The major limitations of this process are the membrane fouling due to accumulation of colloidal substances in the membrane, requirement of pre-treatment procedures like chemical clarification and filtration, membrane scaling due to presence of iron and manganese in the wastewater [57].

Both ultrafiltration and reverse osmosis separate the compounds through a permeable membrane using selective pressure as the driving force. The cut off size for

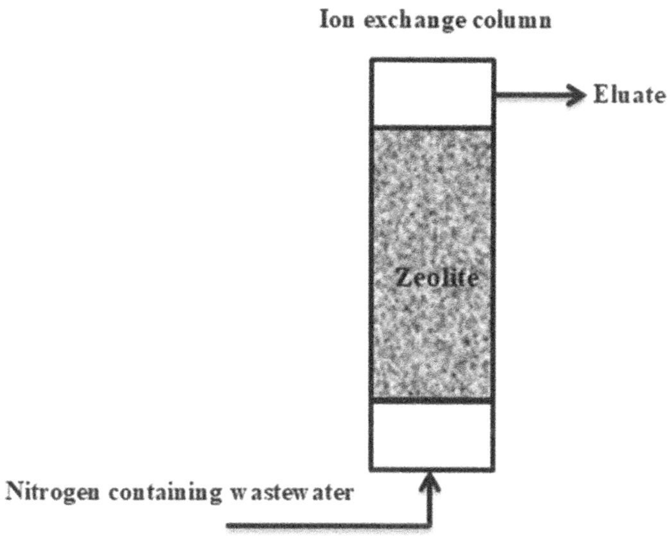

Fig. 3 Nitrogen adsorption through zeolite column

the membrane used in RO is <0.05 μm whereas ultrafiltration is usually used to separate solutes above a molecular weight of 500–2000, which implies a relatively small osmotic pressure, reverse osmosis is used to remove material of low molecular weight, which causes a high osmotic pressure. RO processes are selective for water and small ions whereas the UF separation is based on the molecular size and shape of the molecule. Table 1 summarizes the characteristics of the various membrane separation processes.

Table 1 Membrane separation processes

Membrane process	Working principle	Membrane size range	Function
Reverse osmosis	Pressure gradient	<0.05	Selective transport of water and ions
Ultrafiltration	Pressure gradient	5×10^{-3}–10	Selective to molecular size and shape
Electrodialysis	Electrical potential gradient	<0.1	Selective to ions
Dialysis	Concentration gradient	<0.1	Selective to solute

3.4 Methods of Chemical Precipitation for Nitrogen Removal

Another common method of nitrogen removal from wastewater is by precipitating ammoniacal nitrogen as Magnesium Ammonium Phosphate (MAP), which is also known as struvite. This method has been widely explored and applied for treatment of different types of nitrogen rich wastewaters, such as tannery effluents of leather industries, digester supernatant and in industrial wastewater treatment plants, from coke plants and nitrogen works to pig farms, sludge liquor and landfill leachate.

Commercial magnesium salts and other magnesium containing by-products are used as the chemical precipitating agents. Low-cost Mg sources, such as magnesite, brucite, a by-product of marine salt manufacturing and of the thermal treatment of meat waste, bone meal, etc., have been assessed for its precipitating potential in the formation of struvite [58]. The reaction for struvite formation id given as:

$$Mg^{2+} + PO^{3-} + NH_4{}^{+} + H_2O \rightarrow Mg\,NH_4PO_46H_2O \downarrow$$

Struvite or Magnesium Ammonium Phosphate (MAP) is a white insoluble crystalline compound comprising equivalent concentration of magnesium, ammonium, and phosphate [59]. Though, it is completely accepted that this is a potential method, the usage of costly magnesium salts increases the operational costs, which is a major limitation of this process [60]. However, using inexpensive sources containing MgO or $Mg(OH)_2$ and H_3PO_4 can greatly reduce the operational cost. For example, ninety percent of ammonia was removed using MgO as the coagulant during the treatment of cochineal processed water. In addition, high ammonia removal was achieved during Coking effluent treatment by using pyrolysate of magnesite as the Mg source.

3.5 Air Stripping for Nitrogen Removal

Air stripping is conventional method used for the removal of volatile gases such as hydrogen sulfide, hydrogen cyanide and ammonia by driving air through the waste water. In this method, when alkali is added to the wastewater, non-volatile ammonium ions ($NH_4{}^{+}$) convert to volatile ammonia molecules (NH_3), which can be stripped from the water phase into the gas phase when air is aerated into the water. Then, by aerating the NH_3 containing air into an acid solution, the NH_3 in the gas phase can be absorbed and recovered. The related chemical equilibriums include: The advantages of air stripping lie in its modest reagent costs and easy operation and in the possibility of recovering the removed nitrogen as ammonium sulphate [55]. Air stripping is considered as the most attractive for the treatment of the high nitrogen containing effluent since the cost of one kg of ammonia removal is significantly lower at high ammonia concentrations.

The process of ammonia stripping generally involves the addition of lime or caustic chemical in order to increase the pH of the effluent nearly from 10.8 to 11.5

during which ammonium hydroxide ions reacts to the liberation of ammonia gas according to following equilibrium reaction:

$$NH_4^+ + OH^- \rightarrow H_2O + NH_3 \uparrow$$

The rate of ammonia removal in this method stripping is highly dependent on pH, because the reaction between the ammonium ion and the ammonia gas is an acid base reaction. Therefore, the adjusting the pH of effluent to pH 11 before stripping results in the high ammonia removal. Some of the limitations associated with ammonia stripping are its inefficiency in cold weather, and the formation of calcium carbonate scale in the air stripping tower [61].

4 Plastics: Future Perspectives and Challenges

The increase in discharge of plastics into environment poses adverse effects on the ecosystem. To combat this, legislations related to development of sustainable treatment process for plastic-waste containing wastewater and technically advanced research and analysis for the better understanding on the insights of plastics have to put on practice in the upcoming years.

In the present era, plastics are considered as the inseparable material in our daily life, because of its usage in various aspects such as packages, pharmaceuticals, construction materials, electronic gadgets etc. However, the life span of plastic in environment is a crucial factor to decide its persistent nature and vulnerability [62, 63]. The concept of green chemistry directs our approach towards the development of next generation eco-friendly and biodegradable plastic products which brings a major revolution in the plastics manufacturing industries. An active participation and support from various interdisciplinary groups including industry, academia and government organisations is well required to develop, promote and commercialize these eco-friendly and bio-based polymer products in the society [64, 65].

4.1 Development of Sustainable Polymers

The wide usage of polymeric materials not only provides good economical benefits to the country. In addition, it also increases the rate of non-biodegradable polymeric waste generation associated with major environmental problems. Therefore, there is an urgent need of developing sustainable process to address this problem using two approaches.

(1) Plastic recycling
(2) Developing bio-based eco-friendly polymeric materials [66].

Sustainable polymers i.e. environmentally safer and renewable polymers composed biologically compatible and biodegradable compounds and to provide satisfactory effects to the users/industries. Biopolymers produced from biological source including microbial and plant based components such as carbohydrate and protein polymer as well as from animal origin like wool, silk, gelatin, and collagen are converted into non-toxic end products. Apart from this, synthetic polymers such as polylactic acid (PLA), PVA, and poly (butylenes succinate) (PBS) are also biodegradable. The emerging environmental issues and the increased rate of petroleum all across the globe is the major reason towards the search of sustainable plastics [65, 67].

4.2 Plastic Recycling

Recycling of plastic waste into useful products could bring solutions to the disposal of hazardous materials in landfill sites and to make industrially acceptable commercial products without the use of non-renewable fossil fuels. The practice on recycling the plastic polymers not only reduces the amount of plastic polymeric waste generations but also provides the advantage on the conservation of energy, non-renewable resources, reduces green house gas emissions, etc. Several recycling strategies for post-consumer plastic waste are being developed. The process of melting, shredding or granulation of waste plastics presents a new transformed material. The conversion of plastic waste into oil was performed using direct liquefaction and pyrolysis techniques [68]. Mechanically and chemically treated plastic wastes were used in cement-composites and other construction materials [69]. The hydrocarbons/fuels and refined chemicals are recovered from plastic waste using catalytic conversion, depolymerization and gasification process. The recycling approach of post-consumer plastics with a potential to reduce huge amount of plastic waste is gaining much attention among the scientific communities and industries. This technology could provide immense benefit to plastic waste management sectors [70].

4.3 Biodegradable Plastics: As an Alternative to the Existing Petrochemical-Based Polymers

Technological advancements in synthetic and petroleum-based plastic polymers have contributed several economical benefits to humans. The petroleum derived and synthetic polymers are highly stable in nature. Hence, they are widely used in various sectors such as agriculture, hotels, food and goods packaging, pharmaceutical industries etc. Despite of consistent stability of the plastics in the environment as well as in the market, it is contributing to major issues like the resistant towards biodegradation, bioaccumulation and demand for petroleum oil and their great resistance to biodegradation. On the other hand, the well known problem in the earth is the exhaustion rate of petroleum reserves. Therefore, it becomes mandatory to search for alternate and renewable natural recourses for the plastic production in future.

Biodegradable polymers are now recognised as a better substitute to the conventional petroleum-based plastics [71, 72]. In this context, several research and developments are being widely carried-out for the synthesis of sustainable plastics. Synthetic polyester-based biodegradable plastics which are produced by using few process modifications in polyethylene polymerization and blending agents have been commercialized in food packaging industry, paper industry and garbage carry-bags. Studies have reported that polylactic acid (PLA) and thermoplastic starch-based polyesters polymers are easily degraded into non-toxic end products i.e. water and carbon dioxide, upon the microbial environment. The agricultural products such as potato, corn and wheat starch are the natural resource for starch-based plastic polymers. Another type of biodegradable polymers are produced by bacteria includes polyhydroxyalkanoates (PHAs) [73]. Because of the mechanical property, biodegradability and biocompatibility of PLA polymers, they were greatly used for insurgical implants, drug delivery support materials since 1980s. Inspite of its advantages, the production cost and the rate of biodegrdation of PLA is the major drawback in extensive usage [74].

Starch-based polymers are considered as comparatively low-cost polymers that are produced from agricultural resources. However, the consumption of water in the manufacturing process is high and also influences the strength of the polymeric products. Cellulosic polymers is the another biodegradable material but cannot afford thermal stability. Attempts are made to improve the thermostability and thermoplasticity of cellulosic polymers which in-turn reduces the rate of biodegradability. Henceforth, technical problems are to be resolved to balance the stability and biodegradability of agricultural based polymeric materials. In which, molecular mass and water percentage are the crucial factor for balancing the material property [71, 75]. The synthetic hydrophobic polymers are manufactured and made available since 1990s. Unfortunately the commercially available synthetic hydrophobic polymers are expensive than the petroleum based polymers. Hence, the blending of starch with these polymers has recently become the focus of researchers to reduce the production cost and improves bioaccessibility [71, 76]. The plastic waste obtained from wastewater treatment units can be recycled by using these strategies in order to reduce waste generations and to provide product value in the market [77].

5 Conclusion

Till date, immense resources have been used for the facility development and establishment of wastewater treatment sectors. Deplorably, none have met the standard discharge permits level and operational efficiency. Process modifications in the treatment approaches are needed to provide more reliable performance. Operation and maintenance costs and reliable process control are particularly important to operating agencies. To save energy and resources, consideration towards the selection of process is mandatory. Therefore, to design and operate the efficient waste treatment plant, it is necessary to have depth knowledge of each unit operations involved in the treatment process. Several innovations are made in biological aspects of treating the wastewater. One such process is anammox, which created a huge breakthrough in managing the nitrogen containing wastewater. Other techniques like SHARON, Canon and OLAND are also established to tackle the high strength nitrogen containing wastewater. The finding of extremeophilic microbes i.e. aerobic ammonia oxidizers have paved way for new treatment strategies. In the future the combined process involving NO_x for the conversion of nitrogen containing waste and wastewater will improve the nitrogen removal. In the biological process, crucial factors involved in the successful operation of treatment plant are rate kinetics, oxygen transfer, wastewater characteristics, environmental conditions, operational conditions, maintenance of culture and costs. On these considerations, it is well evident that there is no single best treatment process for nitrogen containing waste present in the municipal landfill leachate or any other kind of wastewater. The new novel and innovative recycling opportunities for the post consumer plastics or nitrogen containing polymers could reduce the disposal rate of complex polymers and can conserve non-renewable resources.

References

1. Xanthos, D., Walker, T.R.: International policies to reduce plastic marine pollution from single-use plastics (plastic bags and microbeads): a review. Mar. Pollut. Bull. (2017). https://doi.org/10.1016/j.marpolbul.2017.02.048
2. Adrados, A., De, M.I., Caballero, B.M., et al.: Pyrolysis of plastic packaging waste: a comparison of plastic residuals from material recovery facilities with simulated plastic waste. Waste Manag. **32**, 826–832 (2012). https://doi.org/10.1016/j.wasman.2011.06.016
3. Geyer, R., Jambeck, J.R., Law, K.L.: Production, use, and fate of all plastics ever made. Sci. Adv., 25–29 (2017)
4. Singh, B., Sharma, N.: Mechanistic implications of plastic degradation. Polym. Degrad. Stab. **93**(3), 561–584 (2008). https://doi.org/10.1016/j.polymdegradstab.2007.11.008
5. Kole, P.J., Löhr, A.J., Van Belleghem, F.G.A.J., Ragas, A.M.J.: Wear and tear of tyres: a stealthy source of microplastics in the environment. Int. J. Environ. Res. Publ. Health (2017). https://doi.org/10.3390/ijerph14101265
6. Watanabe, M., Nakata, C., Wu, W., et al.: Characterization of semi-volatile organic compounds emitted during heating of nitrogen-containing plastics at low temperature. Chemosphere **68**, 2063–2072 (2007). https://doi.org/10.1016/j.chemosphere.2007.02.022
7. He, Z., Li, G., Chen, J., et al.: Pollution characteristics and health risk assessment of volatile organic compounds emitted from different plastic solid waste recycling workshops. Environ. Int. **77**, 85–94 (2015). https://doi.org/10.1016/j.envint.2015.01.004
8. Ghaffar, A., Scott, A., Scott, G.: Initiation process in the photodegradation of high impact polystyrene. Eur. Polym. J. **12**, 615–620 (1976)
9. Tyler, D.R.: Mechanistic aspects of the effects of stress on the rates of photochemical degradation reactions in polymers. J. Macromol. Sci. Polym. Rev. **44**, 351–388 (2004). https://doi.org/10.1081/MC-200033682

10. Sekine, Y.: Oxidative decomposition of formaldehyde by metal oxides at room temperature. Atmos. Environ. **36**, 5543–5547 (2002)
11. Palmisano, A.C., Pettigrew, C.A.: Biodegradability of plastics. Consistent methods for testing claims of biodegradability need to be developed. Am. Inst. Biol. Sci. **42**, 680–685 (2012)
12. Ishigaki, T., Kawagoshi, Y., Ike, M., Fujita, M.: Biodegradation of a polyvinyl alcohol-starch blend plastic film. World J. Microbiol. Biotechnol. **15**(3), 321–327 (1999)
13. Ulbrich, K.: Biodegradation of biomedical polymers. Polym. Sci. **9**, 1–58 (1983)
14. Dave, H., Rao, P.V.C., Desai, J.D.: Biodegradation of starch polyethylene LMS in soil and by microbial cultures. World J. Microbiol. Biotechnol. **13**, 655–658 (1997)
15. Bergmann, M.: Marine Anthropogenic Litter (2015)
16. Yoon, J., Jin, H., Kim, C., Kim, M.: Theoretical prediction of weight loss and molecular weight during random chain scission degradation of polymers. Polymer **38**, 3573–3579 (1997)
17. Haines, J.R., Alexander, M.: Microbial degradation of high molecular weight alkanes. J. Appl. Microbiol. **28**, 1084–1085 (1974)
18. Yang, H., Yoon, J., Kim, M.: Dependence of biodegradability of plastics in compost on the shape of specimens. Polym. Degrad. Stab. **87**, 131–135 (2005). https://doi.org/10.1016/j.polymdegradstab.2004.07.016
19. Holland, K.A., Rae, I.D.: H3cxcH3. Aust. J. Chem. **38**, 17–22 (1987)
20. Czerný, J.: Thermo-oxidative and photo-oxidative aging of polypropylene under simultaneous tensile stress. J. Appl. Polym. Sci. **16**, 2623–2632 (1972). https://doi.org/10.1002/app.1972.070161015
21. Slovenica, A.C., Hrenovi, J.: Biodegradation of Plastic Compost Bags Under Controlled Soil Conditions (2015)
22. Nagai, Y., Ogawa, T., Zhen, L.Y., et al.: Analysis of weathering of thermoplastic polyester elastomers—I. Polyether-polyester elastomers. Polym. Degrad. Stab. **56**, 115–121 (1997). https://doi.org/10.1016/S0141-3910(96)00189-9
23. van Niel, E.W.J., Arts, P.A.M., Wesselink, B.J., et al.: Competition between heterotrophic and autotrophic nitrifiers for ammonia in chemostat cultures. FEMS Microbiol. Lett. **102**, 109–118 (1993). https://doi.org/10.1016/0378-1097(93)90006-N
24. Castignetti, D., Hollocher, T.C.: Heterotrophic nitrification among denitrifiers. Appl. Environ. Microbiol. **47**, 620–623 (1984)
25. Ralt, D., Gomez, R.F., Tannenbaum, S.R.: Conversion of acetohydroxamate and hydroxylamine to nitrite by intestinal microorganisms. Eur. J. Appl. Microbiol. Biotechnol. **12**, 226–230 (1981). https://doi.org/10.1007/BF00499492
26. Robertson, L.A., Kuenen, J.G.: Thiosphaera Pantotropha and Other Bacteria. Antonie Van Leeuwenhoek, pp. 139–152 (1990)
27. Spiller, H., Dietsch, E., Kessler, E.: Intracellular appearance of nitrite and nitrate in nitrogen-starved cells of Ankistrodesmus braunii. Planta **129**, 175–181 (1976). https://doi.org/10.1007/BF00390025
28. Stams, A.J.M., Marinka Flameling, E., Marnette, E.C.L.: The importance of autotrophic versus heterotrophic oxidation of atmospheric ammonium in forest ecosystems with acid soil. FEMS Microbiol. Lett. **74**, 337–344 (1990). https://doi.org/10.1016/0378-1097(90)90686-K
29. Lesley et al.: Heterotrophic Nitrification, pp. 857–863 (1988)
30. Shoun, H., Tanimoto, T.: Denitrification by the fungus Fusarium oxysporum and involvement of cytochrome P-450 in the respiratory nitrite reduction. J. Biol. Chem. **266**, 11078–11082 (1991)
31. Zumft, W.G.: Cell biology and molecular basis of denitrification. Microbiol. Mol. Biol. Rev. **61**, 533–616 (1997)
32. Focht, D.D., Chang, A.C.: Nitrification and denitrification processes related to waste water treatment. Adv. Appl. Microbiol. **19**, 153–186 (1975). https://doi.org/10.1016/S0065-2164(08)70428-3
33. Zumft, W.G.: M06T3102M019N004 **14405**, 43–58 (1997)
34. Zumft, W.G., Viebrock, A., Blümle, S., et al.: Molecular biology of respiratory nitrate reduction (denitrification) of Pseudomonas stutzeri. Inorg. Nitrogen Plants Microorg. **000**, 320–326 (1990). https://doi.org/10.1007/978-3-642-75812-6_48

35. Korner, H., Zumft, W.G.: Expression of denitrification enzymes in response to the dissolved oxygen levels and respiratory substrate in continuous culture of Pseudomonas stutzeri. Appl. Environ. Microbiol. **55**, 1670–1676 (1989)
36. Hellinga, C., Schellen, A.A.J.C., Mulder, J.W., et al.: The SHARON process: an innovative method for nitrogen removal from ammonium-rich waste water. Water Sci. Technol. **37**, 135–142 (1998). https://doi.org/10.1016/S0273-1223(98)00281-9
37. Hellinga, C., Van Loosdrecht, M.C.M., Heijnen, J.J.: Mathematical and Computer Modelling of Dynamical Systems: Methods, Tools and Applications in Engineering and Related Sciences Model Based Design of a Novel Process for Nitrogen Removal from Concentrated Flows, pp. 37–41 (2010)
38. Mulder, J.W., Van Loosdrecht, M.C.M., Hellinga, C., Van Kempen, R.: Full-scale application of the SHARON process for treatment of rejection water of digested sludge dewatering. Water Sci. Technol. **43**, 127–134 (2001). https://doi.org/10.2166/wst.2001.0675
39. Strous, M., Van Gerven, E., Zheng, P., et al.: Ammonium removal from concentrated waste streams with the anaerobic ammonium oxidation (ANAMMOX) process in different reactor configurations. Water Res. **31**, 1955–1962 (1997). https://doi.org/10.1016/S0043-1354(97)00055-9
40. Jetten, M.S.M., Schmid, M., Schmidt, I., et al.: Improved nitrogen removal by application of new nitrogen-cycle bacteria. Rev. Environ. Sci. Biotechnol. **1**, 51–63 (2002). https://doi.org/10.1023/A:1015191724542
41. Schmidt, I., Sliekers, O., Schmid, M., et al.: New concepts of microbial treatment processes for the nitrogen removal in wastewater. FEMS Microbiol. Rev. **27**, 481–492 (2003). https://doi.org/10.1016/S0168-6445(03)00039-1
42. Van Dongen, U., Jetten, M.S.M., Van Loosdrecht, M.C.M.: Ammonium rich wastewater. Growth Lakel **44**, 153–160 (2001)
43. Strous, M., Kuenen, J.G., Jetten, M.S.M.: Key physiology of anaerobic ammonium oxidation. Appl. Environ. Microbiol. **65**, 3248–3250 (1999)
44. Van De Graaf, A.A., De Bruijn, P., Robertson, L.A., et al.: Autotrophic growth of anaerobic ammonium-oxidizing micro-organisms in a fluidized bed reactor. Microbiology **142**, 2187–2196 (1996). https://doi.org/10.1099/13500872-142-8-2187
45. Strous, M., Heijnen, J.J., Kuenen, J.G., Jetten, M.S.M.: The sequencing batch reactor as a powerful tool for the study of slowly growing anaerobic ammonium-oxidizing microorganisms. Appl. Microbiol. Biotechnol. **50**, 589–596 (1998). https://doi.org/10.1007/s002530051340
46. Koch, G., Egli, K., Van Der Meer, J.R., Siegrist, H.: Mathematical modeling of autotrophic denitrification in a nitrifying biofilm of a rotating biological contactor. Water Sci. Technol. **41**, 191–198 (2000). https://doi.org/10.2166/wst.2000.0444
47. Third, K.A., Sliekers, A.O., Kuenen, J.G., Jetten, M.S.M.: The CANON system (completely autotrophic nitrogen-removal over nitrite) under ammonium limitation: Interaction and competition between three groups of bacteria. Syst. Appl. Microbiol. **24**, 588–596 (2001). https://doi.org/10.1078/0723-2020-00077
48. Helmer, C., Tromm, C., Hippen, A., et al.: Single stage biological nitrogen removal by nitritation and anaerobic ammonium oxidation in biofilm systems. Water Sci. Technol. **43**, 311–320 (2001). https://doi.org/10.2166/wst.2001.0062
49. Hippen, A., Helmer, C., Kunst, S., et al.: Six years' practical experience with aerobic/anoxic deammonification in biofilm systems. Water Sci. Technol. **44**, 39–48 (2001). https://doi.org/10.2166/wst.2001.0751
50. Siegrist, H., Reithaar, S., Koch, G., Lais, P.: Nitrogen loss in a nitrifying rotating contactor treating ammonium-rich wastewater without organic carbon. Water Sci. Technol. **38**, 241–248 (1998). https://doi.org/10.1016/S0273-1223(98)00698-2
51. Schmidt, I., Zart, D., Bock, E.: Gaseous NO_2 as a regulator for ammonia oxidation of Nitrosomonas eutropha. Antonie Van Leeuwenhoek Int. J. Gen. Mol. Microbiol. **79**, 311–318 (2001). https://doi.org/10.1023/A:1012038314206
52. Kuai, L., Verstraete, W.: Ammonium removal by the oxygen-limited autotrophic nitrification-denitrification system. Appl. Environ. Microbiol. **64**, 4500–4506 (1998)

53. Water Quality Association—WQA: Chloramine Fact Sheet, p. 5. WQA, Lisle (2013)
54. Water Environment Federation—WEF: Design of Municipal Wastewater Treatment Plants—Manual of Practice, vol. 8, 5th edn., p. 2600. McGraw-Hill Professional, New York (2009)
55. Maroneze, M.M., Zepka, L.Q., Vieira, J.G., et al.: Physico-chemical technologies for nitrogen removal from wastewaters: a review. Rev. Ambient E Agua **9**, 445–458 (2014). https://doi.org/10.4136/1980-993X
56. Jorgensen, T.C., Weatherley, L.R.: Ammonia removal from wastewater by ion exchange in the presence of organic contaminants. Water. Res. **37**, 1723–1728 (2003). https://doi.org/10.1016/S0043-1354(02)00571-7
57. Merlo, R., Wong, J., Occiano, V., et al.: Analysis of organic nitrogen removal in municipal wastewater by reverse osmosis. Water Environ. Res. **84**, 588–595 (2012). https://doi.org/10.2175/106143012x13373550426797
58. Huang, H.M., Xiao, X.M., Yang, L.P., Yan, B.: Removal of ammonium as struvite using magnesite as a source of magnesium ions. Water Pract. Technol. **5**(1) (2010) https://doi.org/10.2166/wpt.2010.007
59. Ryu, H.D., Kim, D., Lee, S.I.: Application of struvite precipitation in treating ammonium nitrogen from semiconductor wastewater. J. Hazard. Mater. **156**, 163–169 (2008). https://doi.org/10.1016/j.jhazmat.2007.12.010
60. Zhou, S., Wu, Y.: Improving the prediction of ammonium nitrogen removal through struvite precipitation. Environ. Sci. Pollut. Res. **19**, 347–360 (2012). https://doi.org/10.1007/s11356-011-0520-6
61. Campos, J.C., Moura, D., Costa, A.P., et al.: Evaluation of pH, alkalinity and temperature during air stripping process for ammonia removal from landfill leachate. J. Environ. Sci. Heal. Part A Toxic/Hazard. Subst. Environ. Eng. **48**, 1105–1113 (2013). https://doi.org/10.1080/10934529.2013.774658
62. Jablonski, N.: Scientific American, a division of Nature America, Inc. Sci. Am. **248**, 121–131 (2010). https://doi.org/10.1038/scientificamerican1004-76
63. James, K., Grant, T.: LCA of Degradable Plastic Bags Centre For Design, p. 17 (2012)
64. Amass, W., Amass, A., Tighe, B.: A review of biodegradable polymers: uses, current developments in the synthesis and characterization of biodegradable polyesters, blends of biodegradable polymers and recent advances in biodegradation studies. Polym. Int. **47**, 89–144 (1998). https://doi.org/10.1002/(SICI)1097-0126(1998100)47:2%3c89::AID-PI86%3e3.0.CO;2-F
65. Scott, G., Wiles, D.M.: Programmed-life plastics from polyolefins: a new look at sustainability. Biomacromol **2**, 615–622 (2001). https://doi.org/10.1021/bm010099h
66. He, A., Li, S., Ma, J., Yang, Z.: Environmental friendly polymer materials for sustainable development. Int. J. Polym. Sci. **2014**, 2–4 (2014). https://doi.org/10.1155/2014/107028
67. Gross, R.A., Kalra, B., Horvath, I.T.: AIChE (Am. Inst. Chem. Eng.) J. **99**, 1–36 (1999)
68. Shah, N., Rockwell, J., Huffman, G.P.: Conversion of waste plastic to oil: direct liquefaction versus pyrolysis and hydroprocessing. Energy Fuels **13**, 832–838 (1999). https://doi.org/10.1021/ef9802431
69. Naik, T.R., Singh, S.S., Huber, C.O., Brodersen, B.S.: Use of post-consumer waste plastics in cement-based composites. Cem. Concr. Res. **26**, 1489–1492 (1996). https://doi.org/10.1016/0008-8846(96)00135-4
70. Contat-Rodrigo, L., Ribes-Greus, A.: Mechanical behaviour of biodegradable polyolefins. J. Non-Cryst. Solids **235–237**, 670–676 (1998). https://doi.org/10.1016/S0022-3093(98)00618-8
71. Gerngross, T., Slater, S.: Biopolymers and the environment letter. Science **6**(80–), 319–328 (2000)
72. Mecking, S.: Nature or petrochemistry?—biologically degradable materials. Angew. Chem. Int. Ed. **43**, 1078–1085 (2004). https://doi.org/10.1002/anie.200301655
73. Wang, H., Sun, X., Seib, P.: Properties of poly(lactic acid) blends with various starches as affected by physical aging. J. Appl. Polym. Sci. **90**, 3683–3689 (2003). https://doi.org/10.1002/app.13001
74. Akutsu, Y., Nakajima-Kambe, T., Nomura, N., Nakahara, T.: Purification and properties of a polyester polyurethane-degrading enzyme from Comamonas acidovorans TB-35. Appl. Environ. Microbiol. **64**, 62–67 (1998). https://doi.org/10.5254/1.3547140

75. Fang, J., Fowler, P.: The use of starch and its derivatives as biopolymer sources of packaging materials. J. Food Agric. Environ. **1**, 82–84 (2003)
76. Ramis, X., Cadenato, A., Salla, J.M., et al.: Thermal degradation of polypropylene/starch-based materials with enhanced biodegradability. Polym. Degrad. Stab. **86**, 483–491 (2004). https://doi.org/10.1016/j.polymdegradstab.2004.05.021
77. Chen, T., Huang, X., Pan, M., Jin, S., Peng, S., Fallgren, P.H.: Treatment of coking wastewater by using manganese and magnesium ores. J. Hazard. Mater. **168**(2–3), 843–847 (2009). https://doi.org/10.1016/j.jhazmat.2009.02.101

Soil Microbiological Recycling and the Virome Role in a Hectare Grassland

Gero Benckiser

Abstract Viruses need a host for replication. In a hectare grassland, for example, a huge amount of diverse cells, majorly microbial cells are present, in which viral inserts replicate either lysogenically with the host genome or lytically after reprogramming the host metabolism towards virus particle production. A third replication mode is "chronic extrusion". Thereby single stranded (ss) viral DNA is channelled through the cytoplasmic membrane into the surrounding environment and double stranded DNA (ds) reintegrates in new host genomes. During lytic virus replication the host cell bursts. The produced virus particles, now termed virions, and the cell nutrients are released and potential pro- and eukaryotic hosts are attracted. In a nutrient enriched environment virions find easier a new genome for reintegrating by using their inherent infectivity and autoinduced quorum-sensing. After having entered a new host lysogenic, lytic replication switching starts. During "chronic extrusion" and lytic replication cycles set free virions can absorb epigenetic information, which then may be spread by horizontal gene transfers. Lysogenic, lytic replication switchings, termed viral shunts, are temperate. Host cell multiplication regulator, carefully controlled in nature. Assumed 1% of the wealth of host cells in a hectare grassland burst lytic virus replication mediated a significant influence on the gross primary production (GPP) is inevitable. Epigenetic genetic information exchanges may improve host's health and correspondingly the focus of this chapter lies on virus mediated nutrient side drain estimates, calculated in their importance for a hectare grassland.

Keywords Infectivity · Symbiosis · Viral nutrient side drains · Food web transitions · Soil restructuring

Highlights

- The widely unexplored viral role in nutrient cycling
- Mutualistic co-evolution of virus-host relationships and control measures
- Virus mediated cell nutrient fluxes, estimate for a hectare grassland.

G. Benckiser (✉)
Institute of Applied Microbiology, Justus-Liebig University, Heinrich Buff-Ring 26-32, Giessen, Germany
e-mail: gero.benckiser@umwelt.uni-giessen.de

G. Benckiser (ed.), *Soil and Recycling Management in the Anthropocene Era*, Environmental Science and Engineering,
https://doi.org/10.1007/978-3-030-51886-8_2

1 Viral Shunts in Terrestrial Ecosystems

Biogeographic ecosystem studies largely concentrate on single taxonomic composition, gene functional potentials, and biotic interactions. Related studies exhibit that worldwide about 42% of the host genomes in cellbiomes about 10^{30} highly infective viral sequences co-shape and influence mutualistically, parasitically the cell multiplication in terrestrial ecosystems [14, 57, 90, 107, 131, 136, 190, 191, 265]. In soils most abundant are viral gemomes carrying bacteria, archaea, and protozoa [22, 24, 84]. They are plant supported global drivers of the carbon and nitrogen cycle and their population density, diversity and dynamic is strongly affected by lysogenic, lytic viral replication [244]. Host genome integrated viral inserts may foster host's wellness but are also responsible for estimated 15% of all human cancers [141, 151]. Host genome integrated viral replication relies on a high infectivity potential, on phage-host quorum-sensing autoinducers, and lysis-lysogeny decisions which may lead to cell destruction. The decision lytic replication, which means host cell bursting, is one of the organismal population density control mechanisms and the cell nutrients, concomitantly released, help co-shaping ecosystems [108, 118, 132, 139, 174, 173, 176, 214, 268]. Powerful and technically progressing genome sequencing machines enable exploring occurrence and role of the largely unexplored viromes in aquatic and terrestrial biotopes. We increasingly understand the environmental co-shaping the viral role in, at least we realized virus cannot replicate without a host and thus parasitic viral entities are found virtually in all unicellular and multicellular creatures [4, 65].

A milliliter marine water, a gram sediment may contain 10^7 viral particles, and in a gram soil integrates in the hosting prokaryotic, eukaryotic microbiome at least 10^9 ten times more viral genomes which survive temporarily restricted as virions [123]. Thus, viral genome inserts when they turn into a lytic cycle play a not insignificant role in shaping the microbial populations dynamic in terrestrial environments, in a hectare grassland [121]. In spite of a in microbiota progressing ecology only few metagenomic studies describe the role of the viromes in terrestrial ecosystems and correspondingly our knowledge about virus community structures and their biotechnological potential is limited. From Ebola outbreaks we know that the inherent infectivity of viral inserts is an important tool in organizing parasitic life, which virus concerned is the most primitive, but most abundant and most successful surviving form on the planet [11, 57, 172]. For example the soft rot causing, pectinolytic Pectobacteriaceae, species *Pectobacterium* and *Dickeya* spp., formerly known as *Erwinia* spp., are carrier of necrotrophic viral inserts. This viral host alliance co-evoluted over millions of years and tell us that Pectobacteriaceae infected plant crops and the many other virus infected microbes, commensal gut bacteria, plants, animals, insects, and humans are confronted with persistent viral infections by becoming more resistant to plagues, drought or extreme temperatures [57, 99, 105, 118, 131]. The viral inserts induces in plants *inter alia* the synthesis of volatile substances which are keeping parasitic insects away by attracting pollinating ones and persistently cryptovirus infected white clover has developed more efficient nitrogen-fixing root nodules guaranteeing a sustainable nitrogen availability. Life span and fecundity enhance in Densovirus infected bollworms and a better resistance to biopesticides [166, 194, 195, 257]. Eight new virulent phages isolates from the rhizosphere of the coastal legume *Medicago marina* are hosted by different *Sinorhizobium meliloti* strains and cooperate with the host [55]. Infective ds phages and ss viruses, usually described in terms of their hosts as virophages, archaeal and bacteriophages, fungal, protozoan, algal, plant, animal, human viruses. After lytic viral replication virions are set free into aquatic and terrestrial ecosystems and vagabond outside of the host around by collecting epigenetic information and attach temporarily on soil particles or on organisms by expanding thus their surviving. The reintegrating in a new host genome depends on virus inherent infectivity, located on tailings and enwrapping as transmission electron microscopy (TEM) images of the phages v_B-Bak1, v_B-Bak6, and v_B-Bak10 are showing and their tailings morphology illustrating consistence with members of the *Siphoviridae* family [70]. Also the three-dimensional structure of the lamda bacteriophage RNA-Polymerase complex (light- and dark grey) and the viral protein Lambda-N (red), a small protein that bounds to a double-stranded DNA target is required for inducing conformational change [130]. The Lambda-N end connects to the 2 e RNA-Polymerase parts and the upper end of Lambda-N connects to RNA (orange) and different regulative proteins (yellow, blue, green). In *P. aeruginosa* phages this lamda bacteriophage trick apparently works similarly and forces the RNA-polymerase to read the virus genome [193].

Has a virion infected a host genome viroplasm formation, reverse transcriptase-mediated tropism switching, host cell splicing, a central mechanism of genetic regulation in eukaryotic cells starts and organize the transfer of outside virion collected information [7, 10, 27, 32, 45, 65, 102, 115, 143, 146, 173, 202, 207, 214, 220, 243, 245, 262]. Plasmodesmata or wounds are preferred viral entry ports, facilitating the introduction of viral skills to tailor novel immune system stabilizing functionalities [149, 204]. Viral infection skills genetic engineers and conjugation chemists use for understanding advancingly world's population decimating pandemics as Ebola and concerning the number of virus related publications grew during the last decade [1, 140, 139, 207, 235]. *Inter alia* we learnt that giant viruses of the *Mimiviridae* family or distant relatives integrate and host small ds phages and ss virophages, which regulate the viral population density of giant viruses and comcomitantly the population dynamic of giant viruses hosting cells [163, 238]. The in various virus types stored genetic information spreads by horizontal gene transfers and by virus-bacteria-mycelial-root-networks, soil particles ingesting earthworms, plant puncturing aphids, xylem-sucking glassy-winged sharp shooters (family *Cicadellidae*), brown plant hoppers, family *Culicidae*, attacking each year 2–3 million tons of plant biomass alone across Asian countries, mosquitoes, birds, and all the other worldwide acting organismal species in environments and with modern techniques, such as group specific PCR of viral genes, 16S, 18S rDNA, clustered regularly interspaced short palindromic repeats (CRISPR), metagenomics classification, PhyloChips comprising 60,000 bacterial operational taxonomic units, mass spectrometry and imaging devices coupled with liquid high pressure or gas chromatography has been found out that after viral lysogenic, lytic replication switching, termed viral shunt, a gram soil of

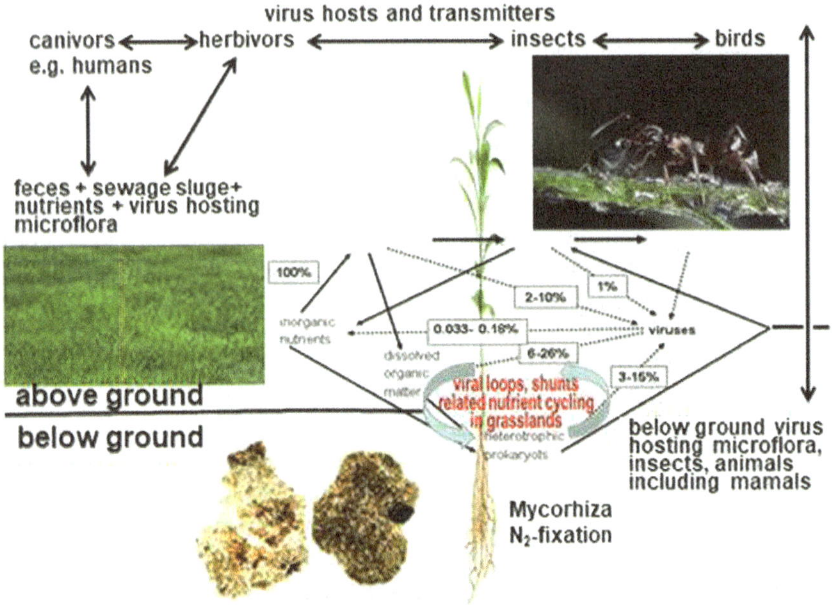

Fig. 1 The environmental virus cycling is of parasitic (deleterious to the host, acute symptoms, chronicity, acquired primarily horizontality), commensal (no direct effect on the host, non-tropism, acquired verticality, horizontality), and mutualistic nature (beneficial to the host and virus, acquired verticality, horizontality). Virus mediated host cell life shortcuts, nutrients and virions releasing from bursting host cells, viral shunts, e.g. in a strongly compartmented grassland ecosystems, influence manifold all types of interrelationships by transferring absorbed outside information spreading it by viral replication, and promoting nutrient cycling, organismal biodiversity and wealth. Adapted from Weinbauer [244], Parker [176]; photo-publication of aphid and ant authorized by Thomas Griessler breschniak@gmx.net)

grass-arable land may contain 1.1 to 4.2×10^9 virions or a handful soil more virions than people on Earth (Fig. 1; [6, 41, 63, 69, 73, 87, 157, 183, 210, 244, 250]).

Barriers for viral host genome reintegration and mutual virus host interrelationships are biofilms, [28] which may counteract the multiplication of prokaryots regulating bacteriovorous communities [3, 77–79, 87, 119, 123, 131, 161, 244]. All these new insights into the viral world, the functioning of viral shunts in relation to their hosts encouraged to estimate in this chapter the possible co-shaping viral influence on a hectare grassland.

2 Microbial and Virus Activities in Terrestrial Ecosystems

A powerful metagenomics analytic helps detecting the soil virome diversity in terrestrial habitats [148, 176, 181, 247]. Virus inserts are key players in the global biogeochemical cycles, define distinct physico-chemical properties of the pro-, eukaryotes species and largely determine the terrestrial microbiome, but our virus concerned knowledge is still a dark matter. With the metagenomics analytic new detected prokaryotic, eukaryotic sequences meanwhile comprise worldwide more than 6.3 million prokaryotic, eukaryotic data bank entries, actually 92 bacteria, 26 archaea, and 5 eukaryotic super groups, all carrying virus inserts [104, 111]. Metagenomics analytic based family trees of all types of organisms can be built, but the sequences of about 16S rRNA sequences of 100 bacteria phyla deposited in the reference databases do not share homology rRNA gene sequences. Through the lack of the metagenomics analytic in physiology explanatory power the physiological input of virus inserts carrying individuals of a microbiome remains hidden. Thus the little understood characteristic multitudes and activity theatres of a large uncultivable number of bacterial, archaeal, protozoal, fungal taxa on skins, mucous membranes, in intestinal tracts, in soils exhibit strong doubts Martiny's statement concerned that across major biomes high proportions of bacteria are culturable [5, 222].

From interpretable, pathogens research related data sets, obtained by combined florescence-in-situ-hybridisation (FISH) and omics-technology, we know that the genome of the anaerobic ammonium oxidizer *Kuenenia stuttgartiensis* consists out of 4,218,325 base pairs and that nitrifying, comammox bacteria and archaea can grow with cyanate and transfer ammonium derived e^- on acceptors as NO_3^- and NO_2^- by reducing them to N_2 or even to NH_3 [58, 103, 127, 226]. New insights into the complexity of the cell metabolism and into the spatial abundance of certain bacteria, archaea, protozoa, and fungi and their interrelationships unveil that around 100,000 unforeseen new antibiotica and enzymes producing bacteria species can be expected among the 10^9 microbial cells in a gram soil [56, 82].

Searching with help of data banks, which store all known RNA molecules present in a cell and all proteins present in cell proteomes, and have the bioinformatics tools for homologies in transcriptomes, and starting at a certain fix point of sequenced ribosomal 16S-RNA of whole genomes (metagenome) then about microbial functions can be speculated, although in data banks are presently more unknown than known proteins deposited and a different amino acid composition of 2% or even one amino acid difference already can mean a complete different protein functioning [18, 36]. Thus, 16S-RNA sequences derived speculations about microbial functions must be handled with care, albeit we are nowadays able to visualize and identify in FISH treated and with stable isotopes spiked soil samples with modern imaging technologies as Raman spectroscopy and NanSIMS (nano-secondary ion mass spectroscopy), characteristic isotope peak changes within a few seconds and can conclude how a cell could chemically be assembled [242, 269].

From the estimated 1.2×10^{30} bacteria and archaea on Earth 4×10^{29} may reside in deep oceanic sub-surfaces, 5×10^{28} in upper oceanic sediments, 3×10^{29} in deep continental sub-surfaces, 3×10^{29} in soils, phyllo-, rhizospheres, and in fewer individuals magnitudes in groundwater, the atmosphere, ocean surface microlayers, on humans and animals skins [79, 78]. All the 1.2×10^{30} bacteria and archaea, involved in earth's matter recycling (Fig. 1), are carrying viral inserts which replicate by using the host metabolism [27]. Besides procaryotes mycoviruses hosting fungi are involved in earth's matter recycling, more than 50,000 one-celled, free-living, viral inserts carrying, bacteria feeding protozoa species per gram soil dry mass, and in a hectare grassland about 2×10^{11} bacteria, archaea feeding nematodes may be found (Fig. 1; [15, 16, 39, 76, 95, 96, 126, 155, 258, 261, 270, 65, 178]). Because across the different phyla of the fungal kingdom mycoviruses seem to share significant similarities, a descenting from a common virus ancestor is suggested and in almost every habitat present parasitic protozoa and nematodes can asymptomatically support host resistance, but also be life threatening to higher organisms. The on bacteria feeding eukaryotic groups control the multiplication of the prokaryotes, their abundance and activity in soils and for example the bacteria grazer *Cephalobus pseudoparvus* decreased, as found by quantitative PCR after 15 and 45 days of soil incubation. The 16S rRNA denitrification genes density by non-significant 60–80% in gene copy numbers, but significantly in the dominant nirK denitrifier community by 8% interestingly not the nirS community activity (Baudoin et al. 2010). In well tempered, pH controlled, with pre-chewed gras fed cow rumen biotopes around 10^5 bacteria, archaea feeding, surplus NH_4^+ excreting protozoa are counted and around 10^5, mostly as spores present fungi (Fig. 3; Table 1; [199]). A cow rumen is dominated by all facets of fermentation and from the available $>10^{10}$ bacteria and arachaea around 7×10^8 are cultivable, but only little understood is how the viral

Table 1 Energy gaining soil microbial metabolic soil energy gaining soil microbial metabolic soil processes

Redox couples		Number of transferred e^-	Eo′ (mV)
CH_2, NH_2, HS, H_2 e^-, H^+		2	−411
Fermentations		1–2	−225–480
CO_2/CH_4		6	−276
SO_4^{2-}/H_2S		8	−214
Fe^{3+}/Fe^{2+}		1	−185
O_2/H_2O		4	+818
NO_3^-/NO_2^-	nirK	2	+432
NO_2^-/NO	nirS	1	+345
NO_2^-/N_2O		4	+818
NO/N_2O		2	+1173
N_2O/N_2	nosZ	2	+1355

Benckiser [22]

inserts carrying protozoa control the prokaryotes in a rumen biotope and contribute to the NH_4^+ availability [168].

In aerated terrestrial ecosystems the organisms prefer to respire with O_2, because the energy gain by oxidizing organic matter is highest. In biofilms [77] and water saturated soil pores [239] only bacteria, archea, fungi and apparently protozoa as the rumen biotope suggests can at O_2 shortage respire with the alternative e^- acceptors CO_2, SO_4^{2-}, Fe^{3+}, NO_3^-, NO_2^-, NO, N_2O and gain energy (adenosinetriphosphate, ATP), while all other has to escape or will die (Table 1; [22]. During photosynthesis H_2O derived e^- are energized, intermittently stored in organic compounds, NH_4^+, CH_4, H_2 and with help of e^- carriers (e.g. NAD^+) returned to H_2O. Methanogens in the cow rumen transfer the e^- and H^+ on CO_2 and CH_4 forms. H_2 respiring Knallgas-bacteria transfer e^- and H^+ on O_2, and fermenting bacteria gain ATP by transferring e^- and H^+ on pyruvate, acetaldehyde and acids are formed (Table 1; [81]). At O_2 shortage bacteria and archaea survive environmental stress better in an exopolymer substance (Eps) surrounded biofilm [27, 79, 78]. Biofilms are similarly sub-structured as soils, in pores, voids, channels in which a distinct O_2 gradient from the outer to the inner biofilm parts favours the electron transport from the donor to the various e^- acceptors as found with modern techniques such as group specific PCR of viral genes, 16S, 18S rDNA, clustered regularly interspaced short palindromic repeats (CRISPR), metagenomics classification, PhyloChips comprising 60,000 bacterial operational taxonomic units, meta-omics technologies, single-cell imaging devices and liquid high pressure or gas chromatography coupled to microscopy, Raman-microspectroscopy, mass spectrometry (Table 1; [38, 3, 30, 87, 119, 127, 175, 131, 267, 225, 135, 145]. EPS surrounded bifilms hosting a variety of bacteria and archaea, fungi, protozoa, provide a remarkable adaptability on pollutants and antibiotics, are barrier for entering viruses, and the available enzyme complexes guarantee surviving [27, 79, 78, 211]. The comammox bacterium *Nitrospira inopinata* forms under varying oxygen regimes with NH_4^+ as e^- and H^+ donator and NO_2^- as e^- and H^+ acceptor the sensitive NO scavenger and N_2O, more than ammonium oxidizing archaea (AOA), but less than ammonium oxidizing bacteria (AOB; [124]).

After the invention of technical N_2 fixation (TNF) by Haber and Bosch mono-culturing farmers started trying to approach nature's high, on biodiversity based biomass productivity and almost doubled the amount of biologically fixed N inputs (BNF). BNF input doubling guaratees a better food security, but the soil inhabitants, long-term adapted on N shortage must adapt on the plant demand and plant assimilation significantly surpassing 118 millon tonnes of worldwide produced nitrogen, on 171 millon tonnes of partly on arable land spread ammonia [47, 117, 216, 67]. Soil biodiversity members not able to adapt on N fertilizer overloads are assumingly overgrown or may disappear. In addition monoculturing landscapes must annually recycle the N in 3000 millions tonnes of worldwide produced crop residues together with the C, N, P, trace elements, and antibiotics stored in slurries of industrial live-stock farming and in sewage sludge producing wastewater treatment plants, uptaking approximately 330 km^3 wastewater [153, 154, 236]. Monocultured, food security guaranteeing crops are rather inefficient in taking up N from the ammonia, produced between 2014 and 2019 in amounts of 171,433 million tons. From 50 to 80 kg N

ha^{-1} $year^{-1}$ surplus field applied N fertilizer plants incorporate only 30–50% [54, 68, 83, 88, 221, 234]. The not uptaken N stresses soil, water, atmosphere, causes biodiversity changes, and finally impairs human health.

N fertilizer use efficiency (NUE) concerned researcher are challenged in avoiding eutrophication, algal blooms in water bodies, nitrate leaching into groundwaters, and farmers cooperate in minimizing CO_2, N_2O, CH_4 emissions [33, 34, 164, 203, 206]. N overloaded environments stimulate the nitrifying and denitrifying consortia in microbiomes (hotspots), which are overrepresented as soil structure co-shapers, carbon, nitrogen sequestrators, presiliencers towards stressors, pollutants degraders, pathogen decontaminators, plant growth promotors, and atmosphere influencers. For understanding the soil, wastewater treatment plant "regulom" better farmers and waste managers are waiting for new technical approaches about which already the Greek prose writer Xenophon speculated, 2500 years before our calculation of time [171]. The reality of understanding the soil metagenomes as finite resources for soil functioning and plant N use efficiency yet is as the microbiologist Prof. Dr. Michael Schloter must answer at a meeting of the Bavarian Academy of Science, 2018, in Munich on a soil diversity related question that we know not much more than Torsvik et al. [233]. By measuring timely the rebuilding of double DNA in soil Torsvik et al. [233] found after PCR invention in 1985 that melted soil DNA not completely re-associates, what means biodiversity can reach new dimensions. Until the wealth of cooperatively working and against each other fighting bacteria, archaea, fungi, protozoa, nematodes, earthworms, and of all other dependent earth inhabitants (Fig. 1) is not reasonably understood farmers and waste managers have to rely on their up to now made experiences.

From 600 in the central Park of New York taken soil samples the DNA was extracted and the obtained DNA structures compared with the global DNA soil biodiversity data record [72]. The New York data set unveiled that the bacteria, archaea diversity found in the central Park of New York is apparently present in each gram of soil. Meanwhile we know that in periodical anoxic soil spells [239] a switching between oxic and anoxic respiratory metabolism takes place [22] and P availability dependent the on 4 sequentially acting denitrification genes relying nitrification-denitrification machinery reduces nitrate to N_2 by gaining energy (Table 1; [17, 61, 184, 206]). From periodical anoxic soil spells and amplicon sequencing of *napA/narG*, *nirK/nirS*, and *nosZ* genes and from transcripts we got insight in the nitrification-denitrification machinery complex and understand better its "early onset" and the cross-talks in between the various regulatory phenotypes of nitrification-denitrification end over produced NO_2^-, NO, N_2O intermediates finally as N_2. It substantiates that the regulatory strategies observed in individual isolates are also displayed in complex communities, meaning that for identifying active key player organisms a successive sampling is prerequsite (Table 1; [144]). In a N over fertilized, agriculturally used Parabrown earth soil (Fig. 2) as in other diverse agricultural production systems the onsetting of the soil-borne, biophysically and nitrification influenced denitrification process depends on denitrifying bacteria, archaea, and fungi in a spatially, temporally variable aggregate size [27, 78,

Fig. 2 Vertical section through the porous soil matrix of a Tschernosem subsoil (C-horizon), channels co-shaped by roots and earthworms, cast infillings, and iron oxide encrusted root residue of a pseudogley. Adapted from Vogel and Babel [239]

239]. At a luxury N availablity also denitrifying microbes seem to act less economically by not using the high N_2O redox potential as efficient as at N shortage. The emission of N_2O increases (Table 1; [25, 162]). In soil aggregates flowing oxygen determines the development of anoxic aggregate centres, the formation of denitrification hotspots, in which Proteobacteria, Acidobacteria, Actinobacteria, Bacteroidetes, Verrucomicrobia, Planctomycetes, few Cyanobacteria, and Archaea are responsible for NO_3^- reduction [29, 72, 89, 209]. Besides the initial oxygen concentration denitrification depends on nitrification, a tangle of minerals, water, nutrients, gases, plant roots, decaying organic matter and the virus inserts behaviour insides of bacteria, archaea, fungi, bacteria grazing protozoa, nematodes, digging earth worms, all working together in nutrients cycling, plant growth, and emission of N_2O or fight against each other [8, 43, 60, 62, 94, 134, 201, 254, 263, 264]. This complex intercommunication seems reasonably be describable by a simple, hyperbolic, oxygen flux dependent denitrification model for soils [206]. For giving an answer on the question what the life of heterotrophic bacterial, archaeal communities in soil aggregates (aggregate-size classes 0.25–0.05, 4.75–2.00, 2.00–0.25, 0.25–0.05 mm) limits Schlüter et al. [206] inoculated soil with 10^9 facultative anaerobic, nosZ-deficient, denitrifying *Agrobacterium tumefaciens* cells ml^{-1} and found that in aggregated soils N_2O emission concerned the soil structure plays an important role. In aggregated soils reside besides bacteria, archaea and fungi nanoscaled chrysophytes, amoeboid, coccoid, palmelloid, morphologically similar or even indistinguishable flagellates of multiple origin within the Apoikiospumella, Chromulinospumella, Segregatospumella, Cornospumella, Acrispumella genera [129]. Such colourless, water and soil inhabiting, bacterivorous flagellates ingest and digest for example bloom-forming, toxic cyanobacteria (*Microcystis aeruginosa*) and golden alga species (*Poterioochromonas* sp.) at an estimated maximal rate of approximately 0.2–1.2 cells h^{-1} [37, 93]. In addition to flagellates soil inhabiting bacteria feeding nematodes as *Cephalobus pseudoparvus* are important co-shapers of the soil microbial activity and diversity, but their ecosystem services are mostly ignored [20, 264]. First when the soil virus-microbe-invertebrate-plant interactions in their aquatic and agricultural role for monoculturing farmers are reasonably recorded and understood by them an approaching of nature's high plant N demand adapted, food security guaranteeing productivity would achievable by monoculturing, soil tillage, N fertilization, a regulating pesticide application, crop rotations, diverse cover crops, and a soil area adapted keeping of animals. Organic farming as concept refuses the use of industrially produced pesticides, while precision farming employs the global position system (GPS) and artificial consumables as nitrification inhibitors (Chap. Plant Demand Adapted Fertilization in Organic and Precision Farming of this book). Important in both approaches of a more ecofriendly food production approaches is the inclusion of crop residue management, animal manure, sewage sludge and spreading of technical fixed reactive nitrogen (N) species in not plant demand surpassing amounts to achieve tolerable greenhouse gas, nitrous oxide (N_2O) emissions [203].

3 Grassland and Nature's Virus Control Measures

About 10^8 microbes may inhabit a gram grassland rhizosphere and bulk soil and on a green leaf area of 1–6 m^2 intensively used grass-arable land, meadows and cattle pastures, tundra/alpine grasslands, and savanna biotopes, covering a huge area of 15, 9, 8; 16 × 10^6 km^2 of the 121 × 10^6 km^2 total terrestrial land surface respectively (leaf area index, LAI, determined to 10 m^2 m^{-2}), about 6.4 × 10^{14} microbial cells may live [59, 112, 142, 185]. The world plant community invests around 400–800 Tg toxin and feeding deterrent-free phloem-C, about 48% of the total annual photosynthetic activity, for stimulating the co-evoluting mutualistic virus-hosting bacteria-mycorrhiza-insect- relationships, the phyllo-, rhizosphere and bulk soil metabolic activity (Fig. 1; [112, 177, 9, 223]). Plants await from the microbial community to receive in return fixed nitrogen and solubilized inorganic soil components. Temperate and tropical grasslands may have a root area index (RAI) of 43 and 80 m^2 m^{-2}, respectively, a fine root C:N:P ratio of 450:11:1, and RAI is apparently 4 to 8 times larger than LAI [53, 59, 112]. Savannas, temperate, and tundra/alpine grasslands can develop in dependency of the phloem-C root feed charge a total root biomass surface of 21, 14, and 10 × 10^9 km^2, a total fine biomass of 14.9, 13.6 and 7.7 × 10^9 km^2, a living fine biomass of 7.7, 8.5 and 2.3 × 10^9 Mg, a fine root length of 0.91, 1.01 and 0.06 × 10^{15} km, and a live fine root surface area of 63.8, 71.2 and 4.2 × 10^7 km^2, respectively, and the phloem flush into rhizosphere rise within minutes the soil microbial community activity from 1 to 2% to up to 60% [26, 35, 112, 218]. With technical fixed N, sewage sludge, or animal manure fertilization similar growth stimulation effects are achieved, a priming of the soil microbial activity. Nitrogen acquisition in the root zone improves. Viral shunts related death rates, in average of 1–2%, as we know from ocean and medicinal research, has for the survivors, for example the microbial activity, a similar growth stimulating effect, and viral infectivity, the lytic replication rates are nature carefully controlled tolerated [47, 75, 131, 146, 214, 224, 64, 244]. Viral infectivity control measures, keeping the death rate in a range of 1–2% are (a) lysogenic replication, (b) a situation adapted reduction of viral adsorption sites, (c) frequent viral receptors changes, (d) initiation of non-viable infections, (e) virus (phage) progeny reduction, (f) excretions of virus digesting enzymes, antibodies, inflammation inhibiting compounds, (g) host cell wall thickening, (h) an orchestration of muralytic, cell wall degrading enzymes, and supportingly act (i) jasmonate dependent systemic resistance (ISR), (j) immune system adjustment, (k) enslavement as viral host genome inserts or rhizobia are demonstrating, (l) apoptosis, (m) programmed cell death at predictable places and times, (n) virus transmitting insect attack prevention, or (o) autophagy, a constitutive, post-natal survival mechanism. The viral shunt frequency control regulates the cell multiplication but concomitantly contributes overcoming productivity limitation by lacking N and other growth factors, e.g. trace element availability and viral shunts help compensating deficits as monoculturing farmers are trying it with plant demand and soil buffer capacity surpassing fertilizations.

Decides the virus insert to leave the host we know from virus hosting cyanobacteria that the host cell metabolic activity increases by ~75% and host cell sequestered carbon, nitrogen and phosphorous are redirected into virus particle production [169, 212, 230]. Is the virus particle production completed, what is termed lytic viral replication, the host cell bursts and the cell nutrients together with the produced virus particles, now termed virions, are set free into the phyllosphere, intestines, and from place to place differing soil pore systems. In soils with their characteristic draining channels, vughs, planar voids, and soil macro- (radius >0.08 mm), meso- (radius 0.075–0.03 mm), micropores (radius <0.03 mm) the activity of the dwellers enhances after viral shunts as after plant residue or cadaver burying their degradation activity and the set free of epigenetic information collecting a virons starts (Figs. 1 and 2; [9, 42, 131, 203, 239]). The released cell nutrients attract virus host candidates and the virion finds easier a new host, can reintegrate, and transfer into the new host genome the outside received and stored information. Such virus related horizontal gene exchanges photosynthesis genes containing cyanobacteria phages exemplify. In the biological world ~$20 \times 10\,s^{-1}$ the cell metabolism co-shaping genetic information transfer events are assumed in dominance among prokaryotes (bacteria, archaea, cell width 0.25–0.5 μm. Because of the information transfer events frequency occur viral infectivity control measures catastrophic outbreaks with haemorrhagic virus involvement and survivors benefit [48, 97, 142, 192, 224, 3, 86, 110, 119, 131, 139, 197, 198, 202, 212, 228, 240, 243, 248, 257].

A bursting cell may release at each viral shunt event around 350 virions, which must find a new host in a narrow host virus range. Is not a new host found the viron decays by setting free at complete ds phage DNA degradation on average 2325384 C-, 823782 N-, 123636 P-atoms which support the soil metabolism nutritionally, but virus-like particles (VLP) vary independent of the DNA content (high, low) in abundance [107, 116, 119]. In the aerobic epilimnion (2 m) and lower anaerobic hypolimnion (50 m) of Lake Bourget (France), for example, the VLP abundance varied from January to August 2008 between 3.4×10^7 and 8.2×10^7 ml^{-1} [31, 180,

197, 215, 229]. During winter the virus-to-bacterium ratio (VBR) was with 69 highest and during winter the viral decay rate increased by 71% in Lake Bourget and ranged between 0.12 and 0.15 day^{-1}. Lysogenic viral replication in Lake Bourget was 0% (spring/summer), 62% (winter) and varied between 75.4% (April) and 8.3% (June). From the viral shunts events in epilimnion and hypolimnion of Lake Bourget also on bacteria grazing protozoa and nematodes increase in numbers by controlling the microbial multiplication. The excreting of uptaken protein surplus NH_4^+ by protozoa and nemadotes contributes to environment nutrition [15, 16, 244, 270]. From oceans is reported that viral shunts are a significant bacterioplankton nutrient source and after each of such events the C, P content of an uptaking cell may increase from 20 and 0.5 fg to up to 56.5 μg C and 1.4 μg P l^{-1} day^{-1}, respectively [75, 169, 212, 230].

Virus populations consist of double stranded (ds) bacteriophages (families *Siphoviridae*, *Myoiviridae*, *Podoviridae*, *Tectiviridae*), single stranded (ss) bacteriophages (families *Microiviridae*, *Circoviridae*, *Nanoviridae*, *Geminiviridae*) and giant ds viruses with integrated ss virophages from distinct phylogenetic origin [51, 75, 116, 122, 128, 131, 152, 159, 186, 196, 230, 232, 244, 255]. In rice fields found ss bacteriophages very likely originated from plants or manured animal faeces. Virions, released into the soil pore hierarchy at viral shunt events attach onto biological or inorganic soil particles and prolong such surviving. The reported interactive and environmental relevant viral roles in oceans encouraged to estimate viral shunt effects which are conceivable to occur in a hectare grassland.

4 A Hectare Grassland Influencing Viral Nutrient Side Drains

Fertilized meadows of the temperate region, temperate grass- and shrub lands (TGS), short grass steppes (Colorado, Great Plains of North America, Mongolian steppe), and tropical savannahs and grasslands (TSG) cover nearly 40% of the land surface and are meanwhile to three-fifth converted into monocotyls grass resembling cereal fields [19, 46, 85, 112, 125, 170, 177, 182, 205, 249]. The macro-, meso-, micropores of grassland topsoil layers may be in size between 0.5 and 15 cm and are responsible for aeration and capillary water flow. The soil density and porosity in meadows, cattle pastures, or unutilized grassland may be 1.57, 1.42 and 1.17 g cm^{-3} and 0.40, 0.46, and 0.56 cm^3 cm^{-3}, respectively, and provide living space per gram topsoil for about 1.7×10^8 phage carrying, patchy distributed, organic residues recycling, rock material dissolving bacteria and archaea [42, 91, 92, 100, 120, 131, 179–189, 217, 231]. The aboveground, majorly in the topsoil rooting grassland plants yield between 0.6 and 12 t dry matter ha^{-1} grassland and is supplied with nutrients by prokaryotes, on prokaryotes grazing eukaryotes, faecal droppings of grass grazing animals, viral shunts, N fertilization, and water related by rainfalls. The yields are highest in TNF fertilized meadows of the temperate regions [170, 177]. Assumed the waterfilm covered soil pores of a hectare grassland are cylindrically shaped and have a by 2•π•r•(r+h) calculable, continuously to soil structure changes subjected surface area (SA) then the space for with viral shunts living bacteria, archaea, fungi, protozoa, nematodes, and through soils moving animals is huge and receives nutrients from various sources [22, 24, 239]. The rhizosphere-bulk soil continuum with bacteriophage dominated plant rhizosheaths have narrow virus-to-bacteria ratios and there are 1.1 to 4.2×10^9 virions to 1.7×10^8 phage carrying bacteria g^{-1} soil ratio estimates [107, 186]. Further is observed that the genetic virome diversity is higher in terrestrial than in aquatic ecosystems.

A hectare grassland nourish 2 grazing livestock units of about 250 kg over a year, from which each has around 8.2×10^{13} body cells, a muscle cell share of 62%, a bone/carcass percentage of 18.4%, and a 150–190 L large rumen (Fig. 3; [2]). Each grazing livestock unit ingests and pre-chews grassland plants and the pre-chewed material is transported into the rumen, which provides for an optimal functioning ca 7×10^{12} virus inserts carrying bacteria, 10^5 bacteria predating protozoa, and 10^5 fungi, mostly present as spores, or for 7 kg microbes a constant pH and temperature optimized working condition (Fig. 3; [3, 87, 199]). The fresh or dried plant material (hay) receiving rumen microflora helps digesting the incoming food and a certain muscle cell viral shunt frequency adds to the nutrient content used for building up muscle and all other cow cells (Table 2). The not by animal cells absorbed nutrients are excreted, reach via excrements the grassland surface and grass plants benefit. The grass plant biomass consists per gram out of 0.84×10^9 cells with a cell dry weight of about 9.8 ng, a 70–90% water content, a leaf-root ratio of around 50–50%, a CN ratio between 20 and 50 and is surface colonized by about 6.4×10^{14} microbial cells per m^2 [74, 142, 170]. Under the assumption that 2% of the grass plant and muscle cells of the 2 cow units, nourished by a ha grassland annually burst and the nutrients may flow viral shunt mediated into the grassland is an assumption and the real dimension is widely neglected and unexplored. From an estimated hourly viral shunt frequency between 0.012 and 17.9% possible nutrient flows (Table 2) in the grassland environment the 7×10^{17} phage carrying bacteria, protozoa and fungi cells in a rumen, the viral shunt surviving, enterobacteriaceae like, 10^6–10^7 phage carrying bacteria and archaea on a square centimeter grassland phylosphere and root surface, and the 1.7×10^8 phage carrying bacteria and archaea in a gram grassland soil benefit. In addition benefit bacteria predating, NH_4^+ excreting protozoa and nematodes in a hectare grassland from faeces droppings, plant residue and cadaver decay [48, 97, 110, 123, 142, 169, 199, 202, 212, 228, 230, 253, 252].

On plant surfaces, in faeces droppings, and soils residing phages hosting *E. coli* like cells represent in the calculations of Table 2 together with the plant, muscle cell nutrient contents the viral shunt subjected cell nutrient flux sources in a ha grassland, inclusively the soil faunal virus mediated cell bursting and related nutrient fluxes. About 3.3×10^{11} phage genome inserts *carrying E. coli* cells would fill out a gram volume and whole-genome sequencing of *E. coli* phages identified the genera *T4virus*, *Js98virus*, *Felix01 virus*, *T1virus*, and *Rtpvirusi* [138]. The dry weight of an *E. coli* cell is on average around 0.003 ng and an exponentially growing bacterial cell may contain 149 ± 8, 35 ± 2, 12 ± 1 fg C, N, P, respectively, whereas a starving bacterial cell may contain significantly less nutrients, e.g., 39 ± 3, 12 ± 2, 2.3 ± 0.6

Table 2 Nutrient contents in a gram muscle, grass (cereal), *E. coli* cells and viral shunt contributions to the nutrient cycling in a hectare grassland (calculation details see below and in the text)

	Nutrient contents in a gram cells of			Annual virus mediated nutrient flows into grassland			
	cow[1)]	grass, cereal[2)]	*E coli*[3)]	cow[a)]+microbes[b)]	plant[c)]+microbes[d)]	soil	total flow
	mg			kg kg	kg t	kg	t
C	41.8	50	500	94.0 + 25.6	12.5 + 3273.4	12.3	3273.5
O	112.5	-	200	-	-	-	-
H	18.3	-	80	-	-	-	-
N	4.7	2	112.5	4.5 + 3.8	0.5 + 894.1	3.2	894.3
P	1.5	1	30	3.4 + 2.1	0.25 + 246.5	0.9	246.6
S	0.4	0.7	10	0.9 + 0.5	0.18 + 46.8	0.45	46.8
Ca	2.6	3	5	0.6 + 0.3	0.75 + 23.4	0.23	23.4
Mg	0.05	1	5	0.1 + 0.3	0.25 + 23,4	0.23	23.4
K	0.37	10	10	0.8 + 0.5	2.5 + 46.8	0.45	46.8
Fe	0.012	0.19	2	0.03+0.1	0.05 + 9.4	0.09	
9.4			}				
Na	0.26	-			- }		
Cl	0.25	-			-		
F	0.007	-	1.8% of		-		
Zn	0.006	0.005	a gram		0.0013		
Cu	0.00018	0.005	*E* coli		0.0013		
Sn	0.00004	-	cell dry	1.2+2.71	-		
Mn	0.00004	0.04	weight		0.01 } 0.94	0.008	
	0.04						
Ni	0.00002		are trace				
Cr	0.00002		elements				
Co	0.000004		~ 0.18 mg				
Mo	0.00002						
B	-	0.005			0.0013		

[a]The muscle cell nutrient content (NCMC; [2] is estimate basis for the virus mediated soil-animal-input)

[b]The grass/cereal plant cell nutrient content (CNP; Fink 1968; https://www.nrcs.usda.gov/wps/PA…/download?cid.ext=pdf) is estimate basis for the virus mediated soil-plant-input

[c]The *E. coli* cell nutrient content (CNC; [81, 66, 137, 260]) is estimate basis for the virus mediated input of the microbial cells in animal intestines, phyllosphere, and soil

[d]On the assumptions of [a, b, c] the estimating of viral influences on a hectare cattle grazed pasture is based

Ruminant share: a hectare grassland nourishes 2, livestock units (200 kg), both 92 kg bones, 308 kg carcass with 62% muscle cells, a daily 2% assumed virus mediated cell burst (VMBCB) → 308,000 g cow carcass × mg nutrients g^{-1} cell biomass/1000 (g) × 2/100 (2% VMBCB) × 365 days + 7000 g rumen microbes × listed *E* coli cell nutrient content nitrogen varies between 85 and 140 (mean 112.5) mg g^{-1} cell biomass/1000 (g) × 2/100 (2% VMBCB) × 365 days [a]ruminant cell share [b]rumen bacteria share+

Plant share: 2 × 6.25 × 10^6 g (mean annual grass + root yield, 50:50%) × mg cell nutrient content/1000 (g) × 2/100 (VMBCB) + 2 × 30^4 × 10^{14} phyllo-, rhizophere bacteria/ha leaf area [59] × cell C, N, P in 30% exponentially growing, 70% starving cells 149, 39 °C, 35, 12 N, 12, 2.3 P fg $cell^{-1}$, respectively, [241]/10^{15} (10^{15} fg = 1 g); other cell nutrients: above listed *E. coli* values in 3.3 × 10^{11} cells per g volume 0.003 ng/g/10^{12} × (g) × 9/100 (mean of 0.012 to 17.9% virus mediated cell burst h^{-1}) × 24 × 365 plant[c], microbe[d] share+

Soil share: 4.26 × 30^4 g ha^{-1} cattle pasture top soil (1.42 g soil cm^{-3}) × 1.7 × 10^8 soil microbes g^{-1} × 0.3 × 149C, 35 N, 12 P fg $cell^{-1}$ (30% of the cells grow exponential) + 0.7 × 39C, 12 N, 2.3 P fg $cell^{-1}$ (70% of the cells starve; [241] fg = g ha^{-1})/10^{15} g microbial nutrents ha^{-1} (for the other cell nutrients the above *E. coli* values/3.8 exponential growth/starving ratio/1000) × 9/100 cell bursts h^{-1} × 24 × 365 + nutrient release from 2.7x10^9 to 50% not a host finding, and completely decaying virions g^{-1} soil $year^{-1}$ × 12 × 2325384 C, or 14 × 823782 N, or 35 × 123636 P atoms/6.022 × 10^{23}

fg C, N, P, respectively (femtogram, fg = 10^{-15} g) [81, 215, 241]. In a litre cell lysate from exponentially growing bacterial cells are inorganic nutrients (2%), 1087–1825 mg C with a N proportion of 51–86%, and protein-enwrapped virions (4-6%; CN ratio ~ 3.8 ± 0.1; prophage DNA/RNA CN ratio ~2; [160]. The C, N, and P cell contents of soil chyitrids and other fungi may range between 38 and 57, 0.23 and 15, and 0.040 and 5.5% on dry biomass basis, respectively, whereby the basidiomycota may have higher C:P, N:P ratios than the ascomycota fungi and ectomycorrhizal fungi may have significantly lower C:N, N:P ratios than saprotrophs [126, 215, 266]. Reportedly soil bacterioprophages may replicate to 22.1–66.7% lysogenically and dependent on pH, salinity, total dissolved nitrogen, ammonium, and phosphorous availability to 4–50% lytically, whereby the atomic C:N, C:P, N:P ratio of exponentially growing and starving cells may vary between 3.8 ± 0.1 and 9.5 ± 1.0, 35 ± 2 and 178 ± 28, and 6.7 ± 0.3 and 18 ± 3, respectively [45, 102, 215, 245, 262, 254, 228].

Since viral shunt frequency mediated nutrient side drain studies in terrestrial ecosystems as grasslands are not available, in the Table 2 reported estimates are of very preliminary and speculative nature. They rely widely on from ocean studies deduced assumptions and it is intended to give on basis of the reported dimensions an idea which nutritional role viral shunts could play in a hectare grassland. Thereby must be considered that the vast majority of identified cosmopolitan viral groups in oxic and hypoxic environments are strongly habitat-type specific, primarily non-lethal, but possessing multiple infectious viral genomes, and vagabonding around virions distribute to the viral replication strategies across the diverse ecosystems [75, 106, 107, 179, 202]. Such multiple infectious viral genomes inside the same unit can even have important implications for viral pathogenesis, antiviral resistance, and social evolution, even in low-energy systems as arctic hypersaline spring sediments having a low activity rate, a small, 8–29% attenuation of growth and production, but a notable viral impact on microbial mortality is observable [50]. Thus, it is very likely that viral shunts also play a not neglectable role in terrestrial ecosytems and should find inclusion in models dealing with nitrogen [21–26].

5 Viral Shunt Effects and the Nutrient Budget of a Hectare Grassland

Table 2 summarizes the roughly calculated nutrient side drains from bursting prokaryotic, plant, and animal cells, which a hectare grassland soil could enrich nutritionally and derivingly viral shunts can contribute with 4564.24 tons of the 10 major and 8 minor cell nutrients to the overall nutrient recycling. Such amounts of nutrients, returned into matter cycling may play a not insignificant role in manipulating the metabolic activity, especially in the soil aggregate hierarchy of grasslands, in the by small distances from each other separated, water retaining soil pores, on leaf and root surfaces, and in the intestines of the 2 grazing livestock units (Figs. 1, 2 and

3). Starving, especially under N shortage suffering viral shunt survivors have evolutionary adaption strategies developed allowing to switch metabolically immediately from starving to luxurious nutritional conditions and finally bacteria grazing protozoa and nematodes as well as cadavers and plant residue decayers, the belowground biodiversity benefit [35, 42, 172, 101, 113, 116, 119, 216]. BNF and viral shunt supported residue mineralization narrow the organic material C:N ratios (85:1, barley, 82:1, rye, 80:1, wheat, 70:1, oat straw, 60:1, corn stalks, 29:1, pea straw, 26:1, rye cover crop, 25:1, mature alfalfa, 24:1, ideal microbe diet, 21:1, clover, 17:1, manure, 15:1, young alfalfa, 11:1, hairy vetch, root residues (humus), on average 62.7, to a soil microbe C:N ratio between 3 and 10:1 and supports the growth of plants, which prefer N uptake concerned a C:N ratio of 10:1 (https://www.nrcs.usda.gov/wps/PA.../download?cid...ext=pdf; [66, 137, 260]). For producing enough food for 2 cattles on a hectare grassland the British Grassland Society recommends to spread 125 kg N ha^{-1} and a stable $d^{15}N$ and $d^{13}C$ isotope ratio analysis unveils that cattles fed with a (C3) grass-based diet show in comparison to a C4 dietary (maize silage) CN differences in their muscle meat [12, 13]. Viral shunt nutrient fluxes may similarly feedback on soil microbes, plants, and cattle cells as maize varieties.

About 3.3×10^{11} bacteria (*E. coli*) cells fill a gram volume, 3.3×10^{11} bacteria (*E. coli*) cells, having an average cell dry weight of around 0.003 ng, and each cell in the exponential growth phase may contain 149 ± 8 fg C, 35 ± 2 fg N, and 12 ± 1 fg P, at starving 39 ± 3 fg C, 12 ± 2 fg N, 2.3 ± 0.6 fg P (fg = femtogram or 10^{-15} g). At virus mediated cell bursting these nutrients will flow in amounts of around 3273.5, 894.3, 246.6, 46.8, 23.4, 23.4, 46.8, 9.4 tons of easily available carbon, nitrogen, phosphorous, if extrapolated to a hectare grassland, in such an environment together with sulfur, calcium, magnesium, potassium, iron, NaCl plus 0.04 tons of trace elements (fluorine, zinc, copper, stannous, manganese, nickel, chrome, cobalt, molybdenum, boron). The surviving organismal diversity and the between 0.6 (without technically fixed TNF) and to 12 t ha^{-1} (with TNF) yielding plants benefit (Table 2; [170, 241]). To the viral shunt related nutrient flow in the water retaining meso- (radius 0.075–0.03 mm), micro- (radius <0.03 mm), ultramicro- (radius 5–0.1 μm), and cryptopores (radius <0.1 μm), bulk soil aggregate hierarchy decaying plant, animal residues and solubilized inorganic soil components contribute (Figs. 1, 2, and 3; [9, 42, 142, 239]). The made viral shunt related estimates (Table 2) indicate that virus mediated around 7000 times more N may annually circulate through a hectare grassland than the British Grassland Society recommends for sustaining a 12 tons per ha high grass productivity. Grasslands soils store >10% of the global soil C and to the conversion of the C stock phloem exudation, transudation, and viral shunt inputs seem to be a not insignificant co-influencer [35, 42, 107, 177]. Yet apparently to the viral shunt and phloem promoted N_2 fixation for reaching highest grass yields TNF fertilization, manuring, or sewage sludge amendments is needed, very likely because root hair cells (15–17 μm in diameter, 80–1500 μm in length), if not mycorrhized, have difficulties in entering the patchy distributed soil pores, especially the micro-, ultramicro-, and cryptopores. Assumed, viral shunts play a similar important role in the 13,963,743 km^{-2} terrestrial land as in oceans [75] then the 1.5×10^{10} g C set free by lysogenic, lytic viral replication switching into the 93.5570.78 km^2 global grassland (67% of the terrestrial earth surface) as extrapolatable from Table 2 would mean a $8.1 \times 10^{-9}\%$ contribution compared with 1856 Pg C, stored in the terrestrial biosphere, or a $2.0 \times 10^{-8}\%$ contribution compared with the 504 Pg C, fixed in the plant biomass, or a $1.1 \times 10^{-8}\%$ contribution compared with the 1352 Pg C, retained by soil organic matter, or $3.0 \times 10^{-7}\%$ contribution compared with the net primary C production of 56.7 Pg C $year^{-1}$, or a $1.0 \times 10^{-8}\%$ contribution compared with the net growth primary C production (GPP) of 123 ± 8 Pg C $year^{-1}$ [71, 109, 158]. The lysogenic, lytic viral replication switching majorly occurs in the soils, rhizo-, and phyllospheres, in the microbially populated, space restricted, μm sized pore volumes and it appears when compared with the net growth primary C production (GPP) as negligible. In a space restricted pore volume autotrophic and heterotrophic O_2, NO_3^-, NO_2^-, NO, or N_2O respiration produces microbial biomass and CO_2 that co-shapes together with biomass ingesting animals the annual net ecosystem production (NEP), defined as difference between net primary production (NPP) and autotrophic R_A, heterotrophic respiration (R_h) (Table 2; [22, 27, 167, 218]). The annual global NEP, estimated in the year 1990 to about 0.8 Pg C $year^{-1}$ with an predicted increase to about 2.6 Pg C $year^{-1}$ d towards the year 2100, varies significantly between the tropics, northern middle and high latitudes, tropical savannahs and grasslands (TSG) and temperate grass- and shrub lands (TGS) by contributing with maybe 31 or 8.5 Pg C $year^{-1}$ (25.2 or 6.9%, respectively) to the GPP of 123 ± 8 Pg C $year^{-1}$, whereby the below-ground biological biomass (BB), the below-ground net primary production (BNPP), and the stocking rate (hectare per animal unit and month) may play their roles ([21, 49, 71, 256]; IPCC report 2019, https://ipcc.ch/report/srccl). The grassland C stock decreases biological stocking rate dependent and the above-ground stocking rate of grassland should not be higher than 0.7 sheep ha^{-1} [49]. For reaching a grassland adapted equilibrium a reduction of the present sheep stocking rate to 0.7 sheep ha^{-1} would need a time span of about 100 years. Relatedly conceivable is that a returning to a stable sustained soil biodiversity after viral shunt in the soil micropore space may takes time, the more because soil pore structure destabilizing processes need to be included in the speculations (Figs. 1 and 2; [9, 52, 189, 217, 239, 246, 251]).

Thus, organizing a plant demand adequate N, P, K, iron, trace element nutrition by food securing with monocultures and by including the soil biology affords agricultural skills (Table 1; [23, 133]; more details in the following chapters: Soil Fauna Activities in Agricultural Greek Landscapes–Plant Demand Adapted Fertilization in Organic and Precision Farming). Iron concerned the worldwide plant demand is about 1.6×10^{28} g Fe and viral shunts in a hectare grassland may contribute to the annual iron demand with 10^7 g Fe (132.3×10^6 km^2 $\times$ 30 cm $\times$ 4.26×10^9 g $\times$ 9.4×10^6 g Fe ha^{-1} (Table 2) $\times$ 100 ha/km^2). In terrestrial soils iron is abundantly available as Fe^{3+}, but the Fe^{3+} absorbability of plants is limited and needs soil biological or farmer's support. In continuously flooded paddy fields rice plants can be productive by having aerenchym tissue that sustains root respiration. Under nutritional imbalance the phenomenon iron toxicity is observed, meaning the rice plant cannot transport sufficient O_2 into the rhizosphere to keep root respiration running and to control Fe_2O_3 respiring bacteria in forming plant uptakeable Fe^{2+}. Too much uptaken Fe^{2+} can intoxify the plant [23]; more details about paddy soils next chapter: N Fertilization Dependent Bacterial and Archaeal Changes in Paddy Soil). In aerated soils plants exude around 48% of their total photosynthetic activity into rhizosheaths, rhizospheres, meso-, micropores of bulk soils to attract Fe^{3+} solubilizing, Fe^{3+} to Fe^{2+} reducing, and N_2 fixing bacteria and archaea for covering their iron and nitrogen demand. Nature started to develop such process balancing strategies from life's origin onwards by beginning with the formation of simple organic compounds as enzymes in diverse numbers (Table 1; [24, 44, 165]). The first *de novo* RNAs, rod-shaped ribonucleic-protein systems, rRNA resemble viroids as viral inserts carrying photosynthesis and N_2 fixation relevant genes of cyanobacteria are documenting [75, 97, 102, 219, 224, 243, 248, 266]. Such innovative microbial structures as viral inserts, able to switch between lysogenic and lytic replication and to collect epigenetic genetic information, are meanwhile detectable in each cell and may improve there lysogenically the host health or regulate the host population size by host cell bursting by concomitantly supporting nutritionally survivors.

Inter alia on viral shunts tolerating biological nitrogen fixation nature may have developed a sustainable and high biomass productivity with a long term on N shortage adapted, high biodiversity. Farmers try approaching with monocultures a similar productivity what seems to be reachable with TNF invention by Haber and Bosch and breeding success. An increasing research focus on a costly landscape N imbalancing and on therapeutic measures lies on managing for example finding ways against Ebola catastrophes with haemorrhagic viral involvement [11, 47, 156, 208]. Nowadays available models not yet include lysogenic, lytic replicating viral host relationships, the inherent self-adaption capability of viral genome inserts, enabling to improve plant and animal health and nutrition. Nowadays models approach solutions primarily, statistically and based on a few variables for example Ludescher et al. [147].

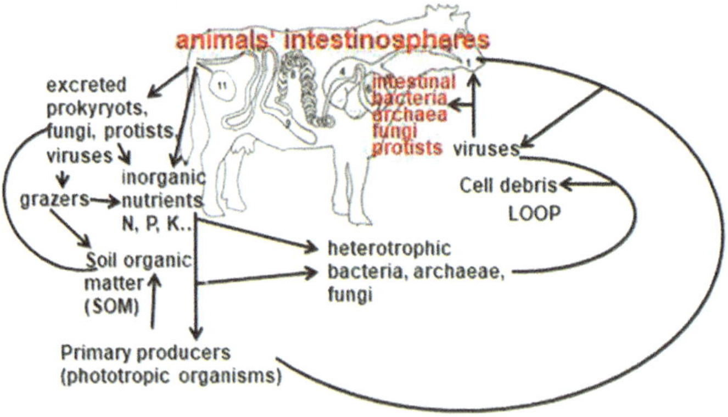

Fig. 3 Nutrient cycling through a grazing ruminant on grasslands

More precise viral shunt forecasting models should be developed and are urgently needed for predicting more accurate feedback loops between viral epigenetic information exchange strategies, molecular structures, exudation/transudation, biotic and abiotic surfaces, and the GPP as largest CO_2 flux driver (Fig. 1; [142, 160, 173, 179, 199, 64, 243, 245, 262]).

6 Conclusions

Viral shunt frequencies in terrestrial ecosystems as grasslands are widely unexplored. For estimating the amount of nutrients with which viral shunts may annually contribute to the nutrient budget of a hectare grassland afforded a couple of assumptions (Table 2) and the made calculations made on this basis exhibit that tons of the 10 major and 8 minor cell nutrients may be channelled viral shunt related through a hectare grassland. Lysogenic, viral replication concerned the publication numbers towards the year 2020 increased explicitly by signalizing viral shunt impacts not only in the terrestrial aggregate hierarchy should play a greater role in nutrient cyling prognosticating models. The intention of this chapter is to inform about the backgrounds and to kick on a broader discussion that may result in a better understanding of viral shunts, exudation/transudation feedback loops, of abiotic soil surface and biotic belowground interrelationships, of molecular, organismal structure changes, in an advancing GPP modelling.

Acknowledgements I thank my former Department Head, the microbiologist Sylvia Schnell, for her great support in publishing this book, the ecologist Stuart Bamforth, Emeritus Tualane University, Department Ecology, Evolution, and Organismal Biology, NewOrleans, Lousiana, USA, and the virologist Prof Dr. Hans-Dieter Klenk, Emeritus Marburg University, Germany for their inputs.

References

1. Abeles, S.R., Pride, D.T.: Molecular bases and role of viruses in the human microbiome. J. Mol. Biol. **426**, 3892–3906 (2014). https://doi.org/10.1016/j.jmb.2014.07.002
2. Aberle, E.D., Forrest, J., Gerrard, D.E, Mills, E.W.: Principles of Meat Science, 4th edn. Kendall/Hunt Publishing Company, Dubuque, Iowa, 354 p, (2001)
3. Abo-Senna, A.S.M.: Occurrence of Bacillus thuringensis bacteriophages in the Egyptian arid soil. Int. J. Virol. Mol. Biol. **6**, 1–8 (2017). https://doi.org/10.5923/j.ijvmb.20170601.01
4. Adachi, A.: Grand challenge in human/animal virology: Unseen smallest replicative entities shape the whole globe. Front Microbiol (2020). https://doi.org/10.3389/fmicrb.2020.00431
5. Amann, R.I., Ludwig, W., Schleifer, K.H.: Phylogenetic identification and in situ detection of individual microbial cells without cultivation. Microbiol. Rev. **59**, 143–169 (1995)
6. Andersen, P.C., Mizell, R.F., Brodbeck, B.V., et al.: Abundance and consumption rate of glassy-winged sharpshooter (Hemiptera: Cicadellidae) on peaches and plums. J. Entomol. Sd. **43,** 394–407. naldc.nal.usda.gov/download/22693/PDF2002 (2008)

7. Andersson, A.F., Banfield, J.F.: Virus population dynamics and acquired virus resistance in natural microbial communities. Science **23**, 1047–1050 (2008). https://doi.org/10.1126/science.1157358
8. Angers, D.A., Eriksen-Hamel, N.S.: Full-inversion tillage and organic carbon distribution in soil profiles: a meta-analysis. SSSAJ **72**, 1370–1374 (2008). https://doi.org/10.2136/sssaj2007.0342
9. Asano, M., Wagai, R.: Evidence of aggregate hierarchy at micro- to submicron scales in an allophanic Andisol. Geoderma **216**, 62–74 (2014). https://doi.org/10.1016/j.geoderma.2013.10.005
10. Ashraf, U., Benoit-Pilven, C., Lacroix, V., Navratil, V., Naffakh, N.: Advances in analyzing virus-induced alterations of host cell splicing. Trends Microbiol. **27**, 268–281 (2018). https://doi.org/10.1016/j.tim.2018.11.004
11. Aurelie, K.K., Guy, M.M., Bona, N.F., Charles, K.M., Mawupemor, A.P., Shixue, L.: A historical review of Ebola outbreaks. An open peer reviewed chapter (2017). https://doi.org/10.5772/intechopen72660
12. Bahar, B., Monahan, F.J., Moloney, A.P., O'Kiely, P., Scrimgeour, C.M., Schmidt, O.: Alteration of the carbon and nitrogen stable isotope composition of beef by substitution of grass silage with maize. Rapid Commun. Mass Spectrom. **19**, 1937–1942 (2005). https://doi.org/10.1002/rcm.2007
13. Bahar, B., Schmidt, O., Moloney, A.P., Scrimgeour, C.M., Begley, I.S., Monaha, F.J.: Seasonal variation in the C, N and S stable isotope composition of retail organic and conventional Irish beef. Food Chem. **106**, 1299–1305 (2008). https://doi.org/10.1016/j.foodchem.2007.07.053
14. Bahram, M., Hildebrand, F., Forslund, S.K., et al.: Structure and function of the global topsoil microbiome. Nature **560**, 233–237 (2018). https://doi.org/10.1038/s41586-018-0386-6
15. Bamforth, S.S.: Protozoa: recyclers and indicators of agroecosystem quality. In: Benckiser, G. (ed.) Fauna in Soil Ecosystems—Recycling Processes, Nutrient Fluxes, and Agricultural Production. Marcel Dekker, New York, USA) (1997)
16. Bamforth, S.S.: Diversity of protozoa. In: Benckiser, G., Schnell, S. (eds.) Biodiversity in agricultural production systems, pp. 205–213. Taylor and Francis, Boca Raton pp (2007)
17. Baraniya, D., Nannipieri, P., Kublik, S., Vestergaard, G., Schloter, M., Schöler, A.: The impact of the diurnal cycle on the microbial transcriptome in the rhizosphere of barley. Microb. Ecol. **75**, 830–833 (2017). https://doi.org/10.1007/s00248-017-1101-0
18. Baric, R.S., Crosson, S., Damania, B., Miller, S.I., Rubin, E.J.: Next-generation high-throughput functional annotation of microbial genomes. mBio, **7**, e01245–16 (2016). https://doi.org/10.1128/mbio.01245-16
19. Battini, F., Agostini, A., Tabaglio, V., Amaducci, S.: Environmental impacts of different dairy farming systems in the Po valley. J. Clean. Product. **112**, 91–102 (2016). https://doi.org/10.1016/j.jclepro.2015.09.062
20. Baudoin, E., Philippot, L., Brauman, A., Villenave, C.: Shifts in size, genetic structure and activity of the soil denitrifier community by nematode grazing. Eur. J. Soil Biol. **46**, 112e118 (2010). https://doi.org/10.1016/j.ejsobi.2009.12.001
21. Beer, C., Reichstein, M., Tomelleri, E., Ciais, P., Jung, M., Carvalhais, N., Rodenbeck, C., Arain, M., Baldocchi, D., Bonan, G.B., Bondeau, A., Cescatti, A., Lasslop, G., Lindroth, A., Lomas, M.R., Luyssaert, S., Margolis, H.A., Oleson, K., Roupsard, O., Veenendaal, E., Viovy, N., Williams, C., Woodward, I., Papale, D.: Terrestrial gross carbon dioxide uptake: global distribution and covariation with climate. Science **329**, 834–838 (2010). https://doi.org/10.1126/science.1184984
22. Benckiser, G.: Organic inputs and soil metabolism. In: Benckiser, G. (ed.) Fauna in Soil Ecosystems—Recycling Processes, Nutrient Fluxes, and Agricultural Production. Marcel Dekker, New York (1997)
23. Benckiser, G., Santiago, S., Neue, H.U., et al.: Effect of fertilization on exudation, dehydrogenase activity, iron reducing populations in the rhizosphere of rice (*Oryza sativa* L.) in relation to iron toxicity. Plant Soil **79**, 305–316 (1984). https://doi.org/10.1007/bf02184324

24. Benckiser, G., Schnell, S.: Biodiversity in agricultural production systems. Taylor and Francis, Boca Raton (2007)
25. Benckiser, G., Syring, K.-M.: Erfassung und Modellierung gasförmiger N Verluste einer Parabraunerde. Z. dt geol. Ges. **136**, 1–6 (1985)
26. Benckiser, G., Ladha, J.K., Wiesler, F.: Climate change and nitrogen turnover in soils and aquatic environments. In: Marxsen, J. (ed.) Climate Change and Microbial Ecology: Current Research and Future Trends (Chap. 8). Caister Academic Press, Norfolk, UK, p. 22
27. Benckiser, G.: Plastics, micro- and nanomaterials, and virus-soil microbe-plant interactions in the environment. In: Prassad, R. (ed.) Plant Nanobionics (Chap. 4). Nanotechnology in the Life Science, pp. 83–96. Springer, Heidelberg (2019)
28. Benckiser, G.: Biofilm surrounded hexachlorobenzene (HCB) crystals and wastewater purification. Examines Mar Biol Oceanogr. **3** (5) (2020). https://doi.org/10.31031/EIMBO.2020.03.000574
29. Bergkemper, F., Kublik, S., Lang, F., Krüger, J., Vestergaard, G., Schloter, M., Schulz, S.: Novel oligonucleotide primers reveal a high diversity of microbes which drive phosphorous turnover in soil. J. Microbiol. Methods **125**, 91–97 (2016). https://doi.org/10.1016/j.mimet.2016.04.011
30. Berry, D., Mader, E., Lee, T.K., Woebken, D., Wang, Y., Zhu, D., Palatinszky, M., Schintlmeister, A., Schmid, M.C., Hanson, B.T., Shterzer, N., Mizrahi, I., Rauch, I., Decker, T., Bocklitz, T., Popp, J., Gibson, C.M., Fowler, P.W., Huang, W.E., Wagner, M.: Tracking heavy water (D2O) incorporation for identifying and sorting active microbial cells. PNAS **112**, E194–E203 (2015). https://doi.org/10.1073/pnas.1420406112
31. Bettarel, Y., Sime-Ngando, T., Amblard, C., Dolan, J.: Viral activity in two contrasting lake ecosystems Appl. Environ. Microbiol. **70**, 2941–2951 (2004). https://doi.org/10.1128/aem.70.5.2941-2951.200
32. Beutin, L., Hammerl, J.A., Strauch, E., et al.: Spread of a distinct Stx2-encoding phage prototype among Escherichia coli O104:H4 strains from outbreaks in Germany, Norway, and Georgia. J. Virol. **86**, 10444–1055 (2012). https://doi.org/10.1128/jvi.00986-12
33. Bhowmik, A., Fortuna, A.-M., Cihacek, L.J., Bary, A.I., Cogger, C.G.: Use of biological indicators of soil health to estimate reactive nitrogen dynamics in long-term organic vegetable and pasture systems. Soil Biol. Biochem. **103**, 308–319 (2016). https://doi.org/10.1016/j.soilbio.2016.09.004
34. Bhowmik, A., Fortuna, A.-M., Cihacek, L.J., Bary, A.I., Cogger, C.G.: Potential carbon sequestration and nitrogen cycling in long-term organic management systems. Renew. Agri. Food. Syst. 1–13 (2017). https://doi.org/10.1017/si742170516000429
35. Blagodatskaya, E., Kuzyakov, Y.: Active microorganisms in soil: critical review of estimation criteria and approaches. Soil Biol. Biochem. **67**, 192–211 (2013). https://doi.org/10.1016/j.soilbio.2013.08.024
36. Bloom, J.D., Arnold, F.H.: In the light of directed evolution: pathways of adaptive protein evolution PNAS 16, 106 (Supplement 1): 9995–10000 (2009). https://doi.org/10.1073/pnas.0901522106
37. Boenigk, J., Pfandl, K., Stadler, P., Chatzinotas, A.: High diversity of the '*Spumella*-like' flagellates: an investigation based on the SSU rRNA gene sequences of isolates from habitats located in six different geographic regions. Environ. Microbiol. **7**, 685–697 (2005). https://doi.org/10.1111/j.1462-2920.2005.00743.x
38. Borriello, G., Werne, E., Roe, F., Kim, A.M., Ehrlich, G.D., Stewart, P.S.: Oxygen limitation contributes to antibiotic tolerance of *Pseudomonas aeruginosa* in biofilms. Antimicrobial agents and chemotherapy 48, 2659–2664 (2004). https://doi.org/10.1128/aac.48.7.2659-2664.2004
39. Bouwman, L.A., Zwart, K.B.: The ecology of bacterivorous protozoans and nematodes in arable soil. Agri. Ecosys. Environm. **51**, 145–160 (1994). https://doi.org/10.1016/0167-8809(94)9004
40. Bratbak, G., Heldal, K.: Significance of algal viruses and ecology of Phaeocystis host-virus interactions. In: Bell, C., Brylinsky, M., Johnson-Green, P. (eds.) Microbial Biosystems. Atlantic Canada Society for Microbial Ecology, Halifax, pp. 571–575 (2000)

41. Brault, V., Uzest, M., Monsion, B., et al.: Aphids as transport devices for plant viruses. Compt. Rend. Biol. **333**, 524–538 (2010). https://doi.org/10.1016/j.crvi.2010.04.001.epub2010
42. De Bruyn, J.: The dead and their microbes. Microbe **11**, 119–124 (2016)
43. Caesar-TonThat, T.C., Lenssen, A.W., Caesar, A.J., Sainju, U.M., Gaskin, J.F.: Effects of tillage on microbial populations associated to soil aggregation in dryland spring wheat system. Eur. J. Soil Biol. **46**, 119 (2010). https://doi.org/10.1016/j.ejsobi.2009.12.004
44. Carvalhais, L.C., Dennis, P.G., Fedoseyenko, D., et al.: Root exudation of sugars, amino acids, and organic acids by maize as affected by nitrogen, phosphorus, potassium, and iron deficiency. J. Plant Nutr. Soil Sci. **174**, 3–11 (2011). https://doi.org/10.1002/jpln.201000085
45. Casjens, S.: Prophages and bacterial genomics: what have we learned so far? Mol. Microbiol. **49**, 277–300 (2003). https://doi.org/10.1046/j.1365-2958.2003.03580.x
46. Cela, S., Ketterings, Q.M., Czymmek, K., et al.: Characterization of nitrogen, phosphorus, and potassium mass balances of dairy farms in New York State. J. Dairy Sci. **97**, 7614–7632 (2014)
47. Chen, J.G., Crooks, R.M., Seefeldt, L.C., Bren, K.L., Bullock, R.M., Darensbourg, Y., Holland, P.L., Hoffman, B., Janik, M.J., Jones, A.K., Kanatzidis, M.G., King, P., Lancaster, M., Lymar, S.V., Pfromm, P., Schneider, W.F., Schrock, R.R.: Beyond fossil fuel–driven nitrogen transformations. Science **360**, 6391 (2018). https://doi.org/10.1126/science.aar6611
48. Chen, H.Y., Di Mascio, M., Perelson, A.S., et al.: Determination of virus burst size in vivo using a single-cycle SIV in rhesus macaques. PNAS **104**, 19079–19084 (2007). https://doi.org/10.1073/pnas.0707449104
49. Chen, Y., Lee, P., Lee, G., Mariko, S., Oikawa, T.: Simulating root responses to grazing of a Mongolian grassland ecosystem Plant Ecology 183, 265–275 (2006). https://doi.org/10.1007/s11258-005-9038-7
50. Colangelo-Lillis, J., Wing, B.A., Raymond-Bouchard, I., Whyte, L.G.: Viral induced microbial mortality in arctic hypersaline spring sediments. Front. Microbiol. (2017). https://doi.org/10.3389/fmicb.2016.02158
51. Colson, P., La Scola, B., Raoult, D.: Viruses of amoebae: a journey through innovative research and paradigm changes. Ann. Rev. Virol. **4**, 61–85 (2017). https://doi.org/10.1146/annurev-virology-101416-04181
52. Corinaldesi, C., Dell'Anno, A., Danovaro, R.: Viral infections stimulate the metabolism and shape prokaryotic assemblages in submarine mud volcanoes. ISME J. **6**, 1250–1259 (2012). https://doi.org/10.1038/ismej.2011.185
53. Coupland, R.T. (ed.): Ecosystems of the World: 8A and B Natural Grasslands. Elsevier, Amsterdam (1993)
54. Craswell, E.T., Godwin, D.C.: The efficiency of nitrogen fertilizers applied to cereals in different climates. Adv. Plant Nutrit. **1**, 1–55 (1984)
55. Cubo, M.T., Alías-Villegas, C., Balsanelli, E., Mesa, D., de Souza, E., Espuny, M.R.: Diversity of *Sinorhizobium (Ensifer) meliloti* bacteriophages in the rhizosphere of *Medicago marina*: Myoviruses, filamentous and N4-Like Podovirus. Front. Microbiol. (2020). https://doi.org/10.3389/fmicb.2020.00022
56. Cycoń, M., Mrozik, A., Piotrowska-Seget, Z.: Antibiotics in the soil environment—degradation and their impact on microbial activity and diversity. Front. Microbiol. **10**, 338 (2019). https://doi.org/10.3389/fmicb.2019.00338
57. Czajkowski, R.: May the phage be with you? Prophage-Like elements in the genomes of soft rot *Pectobacteriaceae*: *Pectobacterium* spp. and *Dickeya* spp. Front. Microbiol. (2019). https://doi.org/10.3389/fmicb.2019.00138
58. Daims, H., Lebedeva, E.V., Pjevac, P., Han, P., Herbold, C., Albertsen, M., Jehmlich, N., Palatinszky, M., Vierheilig, J., Bulaev, A., Kirkegaard, R.H., von Bergen, M., Rattei, T., Bendinger, B., Nielsen, P.H., Wagner, M.: Complete nitrification by Nitrospira bacteria. Nature **528**, 504–509 (2015). https://doi.org/10.1038/nature16461
59. Darvishzadeh, R., Skidmore, A., Schlerf, M., Atzberger, C., Corsi, F., Cho, M.: Estimation of leaf area index and Chlorophyll For a mediterranean grassland using hyperspectral data. In: The International Archives of the Photogrammetry, Remote Sensing and Spatial Information Sciences, vol. 27, Part B7, Beijing (2008)

60. Debnath, A., Reed, S.L.: *Entamoeba histolytica*: from high-throughput technology to new drugs. In: De Pablos Torró, L.M., Morales, J.-L. (eds.) Protozoan Parasitism: From Omics to Prevention and Control. Caister Academic Press, UK, pp. 157–170 (2018). https://doi.org/10.21775/9781910190838;
61. Ding, G., Radl, V., Schloter-Hai, B., Jechalke, S., Heuer, H., Smalla, K., Schloter, M.: Dynamics of soil bacterial communities in response to repeated application of manure containing sulfadiazine. PLoS ONE **9**, e92958. https://doi.org/10.1371/journal.pone.0092958
62. Djigal, D., Baudoin, E., Philippot, L., Brauman, A., Villenave, C.: Shifts in size, genetic structure and activity of the soil denitrifier community by nematode grazing. Eur. J. Soil Biol. **46**, 112e118 (2010). https://doi.org/10.1016/j.ejsobi.2009.12.001
63. Dou, D., Hernandez-Neuta, I., Wang, H., Ostbye, H., Qian, X., Thiele, S., Resa-Infante, P., Kouassi, N.M., Sender, V., Hentrich, K., Mellroth, P., Henriques-Normark, B., Gabriel, G., Nilsson, M., Daniels, R.: Analysis of IAV replication and co-infection dynamics by a versatile RNA viral genome labeling method. Cell Rep. **20**, 251–263 (2017). https://doi.org/10.1016/j.celrep.2017.06.021
64. Van Driesche, R.G., Carruthers, R.I., Center, T., Hoddle, M.S., Hough-Goldstein, J., Morin, L., Smith, L., Wagner, D.L., Blossey, B., BrancatiniV, Casagrande R., Causton, C.E., Coetzee, J.A., Cuda, J., Ding, J., Fowler, S.V., Frank, J.H., Fuester, R., Goolsby, J., Grodowitz, M., Heard, T.A., Hill, M.P., Hoffmann, J.H., Huber, J., Julien, M., Kairo, M.T.K., Kenis, M., Mason, P., Medal, J., Messing, R., Miller, R., Moore, A., Neuenschwander, P., Newman, R., Norambuena, H., Palmer, W.A., Pemberton, R., Perez-Panduro, A., Pratt, P.D., Rayamajhi, M., Salom, S., Sands, D., Schooler, S., Schwarzländer, M., Sheppard, A., Shaw, R., Tipping, P.W., van Klinken, R.D.: Classical biological control for the protection of natural ecosystems. Biol. Control **54**, 2–33 (2010). https://doi.org/10.1016/j.biocontrol.2010.03.003
65. Dávila-Ramos, S., Castelán-Sánchez, H.G., Martínez-Ávila, L., del Rayo Sánchez-Carbente, M., Peralta, R., Hernández-Mendoza, A, Dobson, A.D.W., Gonzalez, R.A., Pastor, N., Batista-García, R.A.: A review on viral metagenomics in extreme environments. Front. Microbiol. (2019). https://doi.org/10.3389/fmicb.2019.02403
66. Eiland, F., Klamer, M., Lind, A.-M., Leth, M., Bååth, E.: Influence of initial C/N ratio on chemical and microbial composition during long term composting of straw. Microbi. Ecol. **41**, 272–280 (2001). https://www.jstor.org/stable/4251820
67. FAO.: The State of Food and Agriculture 2019 (2019). http://www.fao.org/3/CA6030EN/CA6030EN.pdf
68. Fageria, N.K., Baligar, V.C.: Enhancing nitrogen use efficiency in crop plants Adv. Agronomy **88**, 97–185 (2005). https://doi.org/10.1016/S0065-2113(05)88004-6
69. Fang, J.: A world without mosquitoes. Nature **466**, 432–434 (2010). https://doi.org/10.1038/466432a
70. Farlow, J., Bolkvadze, D., Leshkasheli, L., Kusradze, I., Kotorashvili, A., Kotaria, N., Balarjishvili, N., Kvachadze, L., Nikolich, M., Kutateladze, M.: Genomic characterization of three novel Basilisk-like phages infecting *Bacillus anthracis*. BMC Genom. **19**, 685 (2018). https://doi.org/10.1186/s12864-018-5056-4
71. Field, C.B., Raupach, M.R.: The global carbon cycle. Scope Series 62. Island Press, Washington DC (2004)
72. Fierer, N., Jackson, R.B.: The diversity and biogeography of soil bacterial communities. PNAS **103**, 626–631 (2006). https://doi.org/10.1073/pnas.0507535103
73. Filippis, V., Villareal, L.: Lateral gene transfers or viral colonization? Science **293**, 1048 (2001). https://doi.org/10.1126/science.293.5532.1048a
74. Fink, A.: Pflanzenernährung in Stichworten. Hirt-Verlag, Kiel (1969)
75. Finke, J.F., Hunt, B.P.V., Winter, C., Carmack, E.C., Suttle, C.A.: Nutrients and other environmental factors influence virus abundances across oxic and hypoxic marine environments. Viruses **9**, 152 (2017). https://doi.org/10.3390/v9060152
76. Fischer, G.: Diversity of soil fungi. In: Benckiser, G., Schnell, S. (eds.) Biodiversity in Agricultural Production Systems, pp. 149–163. Taylor and Francis, Boca Raton pp (2007)

77. Flemming, H.-C., Wingender, J., Szewzyk, U., Steinberg, P.D., Rice, S.C., Kjelleberg, S.: Biofilms: an emergent form of bacterial life. Nat. Rev. Microbiol. **14**, 563–575 (2016). https://doi.org/10.1038/nrmicro.2016.94
78. Flemming, H.-C., Wuertz, S.: Bacteria and archaea on Earth and their abundance in biofilms. Nat. Rev. Microbiol. **17**, 247–260 (2019). https://doi.org/10.1038/s41579-019-0158-9ature
79. Flemming, H.-C., Tamachkiarowa, A., Klahre, J., Schmitt, J.: Monitoring of fouling and biofouling in technical systems. Water Sci Technol 38, 291–298 (1998). https://doi.org/10.1016/S0273-1223(98)00704-5
80. Forterre, P.: The universal tree of life: an update. Front. Microbiol. **6**, 717 (2015). https://doi.org/10.3389/fmicb.2015.00717
81. Fuchs, G.: Allgemeine Mikrobiologie. Georg Thieme Verlag, Germany (2007)
82. Gans, J., Wolinsky, M., Dunbar, J.: Computational improvements reveal great bacterial diversity and high metal toxicity in soil. Science **309**, 1387–1390 (2005). https://doi.org/10.1126/science.1112665
83. Garnett, T., Conn, V., Kaiser, B.N.: Root based approaches to improving nitrogen use efficiency in plants Plant. Cell Environ. **32**, 1272–1283 (2009). https://doi.org/10.1111/j.1365-3040.2009.02011.x
84. Garrett, R.A., Klenk, H.-P. (eds.): Archaea: evolution, physiology, and molecular biology (2007). Online ISBN: 9780470750865, Blackwell Publishing Ltd. https://doi.org/10.1002/9780470750865
85. Giarrizzo, E., Burrascano, S., Chiti, T., de Bello, F., Lepš, J., Zavattero, L., Blasi, C.: Re-visiting historical semi-natural grasslands in the Apennines to assess patterns of changes in species composition and functional traits Appl. Veget. Sci. **20**, 247–258 (2017). https://doi.org/10.1111/avsc.12288
86. Gilbert, C., Peccoud, J., Chateigner, A., Moumen, B., Cordaux, R., Herniou, E.A.: Continuous influx of genetic material from host to virus populations. PLoS Genet. **12**, e1005838 (2016). https://doi.org/10.1371/journal.pgen.100583
87. Gillis, A., Mahillon, J.: Phages preying on *Bacillus anthracis*, *Bacillus cereus*, and *Bacillus thuringiensis*: past, present and future. Viruses **6**, 2623–2672 (2014). https://doi.org/10.3390/v6072623
88. Good, A.G., Beatty, P.H.: Fertilizing nature: a tragedy of excess in the commons. PLoS Biol. **9**, e1001124 (2011). https://doi.org/10.1371/journal.pbio.1001124
89. Grafe, M., Goers, M., von Tucher, S., Baum, C., Zimmer, D., Leinweber, P., Vestergaard, G., Kublik, S., Schloter, M., Schulz, S.: Bacterial potentials for uptake, solubilization and mineralization of extracellular phosphorus in agricultural soils are highly stable under different fertilization regimes. Environment. Microbiol. Rep. **10**(3): 320–327 (2018). https://doi.org/10.1111/1758-2229.12651
90. Graham, E.B., Paez-Espino, D., Brislawn, C., Neches, R.Y., Hofmockel, K.S., Wu, R., et al.: Untapped viral diversity in global soil metagenomes. biorxiv. [preprint] (2019). https://doi.org/10.1101/583997
91. Grayston, S.J., Griffith, G.S., Mawdsley, J.L., Campbell, C.D., Bardgett, R.D.: Accounting for variability in soil microbial communities of temperate upland grassland ecosystems. Soil Biol. Biochem. **33**, 533–551 (2001). https://doi.org/10.1016/s0038-0717(00)00194-2
92. Griffiths, R.I., Whiteley, A.S., O'Donnell, A.S., Bailey, M.J.: Influence of depth and sampling time on bacterial community structure in an upland grassland soil. FEMS Microbiol. Ecol. **43**, 35–43 (2003). https://doi.org/10.1111/j.1574-6941.2003.tb01043.x
93. Grossmann, L., Bock, C., Schweikert, M., Boenigk, J.: Small but manifold-hidden diversity in "Spumella-like flagellates". J. Eukaryot Microbiol. **63**, 419–439 (2016). https://doi.org/10.1111/jeu.12287
94. Grujcic, V., Nuy, J.K., Salcher, M.M., Shabarova, T., Kasalicky, V., Boenigk, J., Jensen, M., Simek, K.: Cryptophyta as major bacterivores in freshwater summer plankton. ISME J. **12**, 1668–1681 (2018). https://doi.org/10.1038/s41396-018-0057-5)
95. Hartmann, A., Smalla, K., Soerensen, J.: Microbial diversity in the rhizosphere: methodology to study plant-beneficial bacteria. In: Benckiser, G., Schnell, S. (eds.) Biodiversity in Agricultural Production Systems, pp. 101–131. Taylor and Francis, Boca Raton pp (2007)

96. Hayes, S., Mahony, J., Nauta, A., Van Sinderen, D.: Metagenomic approaches to assess bacteriophages in various environmental niches. Viruses **9**, E127 (2017). https://doi.org/10.3390/v9060127
97. Hellweger, F.L.: Carrying photosynthesis genes increases ecological fitness of cyanophage in silico. Environ. Microbiol. **11**, 1386–1394 (2009). https://doi.org/10.1111/j.1462-2920.2009.01866.x. Epub 2009 Jan 23
98. Herzog, S., Wemheuer, F., Wemheuer, B., Daniel, R.: Effects of fertilization and sampling time on composition and diversity of entire and active bacterial communities in German grassland soils. PLoS ONE **10**, e0145575 (2015). https://doi.org/10.1371/journal.pone.0145575
99. Hillesland, K.L., Stahl, D.A., Harcombe, W.: Measuring the cost of microbial mutualism. Microbe **6**, 427–434 (2011)
100. Hirmas, D.R., Gimenez, D., Nemes, A., et al.: Climate—induced changes in continental—scale soil macroporosity may intensify water cycle. Nature **651**, 100–103 (2018). https://doi.org/10.1038/s41596.018.0463.x
101. Hobbie, J.E., Hobbie, E.A.: Microbes in nature are limited by carbon and energy: the starving-survival lifestyle in soil and consequences for estimating microbial rates. Front. Microbiol. (2013). https://doi.org/10.3389/fmicb.2013.00324
102. Howard-Varona, C., Hargreaves, K.R., Abedon, S.T., Sullivan, M.B.: Lysogeny in nature: mechanisms, impact and ecology of temperate phages. ISME J. **11**, 1511–1520 (2017). https://doi.org/10.1038/ismej.2017.16
103. Hu, Z., Wessels, H.J.C.T., van Alen, T., Jetten, M.S.M., Kartal, B.: Nitric oxide-dependent anaerobic ammonium oxidation. Nat. Commum. **10**, 1244 (2019). https://doi.org/10.1038/s41467-019-09268-w
104. Hug, L.A., Baker, B.J., Anantharaman, K., Brown, C.T., Probst, A.J., Castelle, C.J., Butterfield, C.N., Hernsdorf, A., Amano, Y., Ise, K., Suzuki, Y., Dudek, N., Relman Finstad, K.M., Amundson, R., Thomas, B.C., Banfield, J.F.: A new view of the tree of life. Nat. Microbiol. *1*, 16048 (2016). https://doi.org/10.1038/nmicrobiol.2016.48
105. Humphrey, P.T., Nguyen, T.T., Villalobos, M.M., Whiteman, N.K.: Diversity and abundance of phyllosphere bacteria are linked to insect herbivory. Mol. Ecol. **23**, 1497–1515 (2014). https://doi.org/10.1111/mec.12657
106. Humphreys, I.R., Sebastian, S.: Novel viral vectors in infectious diseases. Immunology **153**, 1–9 (2018). https://doi.org/10.1111/imm.12829
107. Hyman, P., Abedon, S.T. (eds.): Viruses of Microorganisms. Caister Academic Press (2018). https://doi.org/10.21775/9781910190852
108. Igler, C., Abedon, S.T.: Commentary: a host-produced quorum-sensing autoinducer controls a phage lysis-lysogeny decision. Front. Microbiol. **10**, 1171 (2019). https://doi.org/10.3389/fmicb.2019.01171
109. Ito, A., Oikawa, T.: Global mapping of terrestrial primary productivity and light-use efficiency with a process-based model. In: Shiyomi, M., et al. (eds.), Global Environmental Change in the Ocean and on Land, TERRAPUB, pp. 343–358 (2004)
110. Iwami, S., Holder, B.P., Beauchemin, C.A.A., et al.: Quantification system for the viral dynamics of a highly pathogenic simian/human immunodeficiency virus based on an in vitro experiment and a mathematical model. Retrovirology **9**, 18 (2012). https://doi.org/10.1186/1742-4690-9-18
111. Jackson, W.T., Coyne, C.B. (eds.): Enteroviruses: Omics, Molecular Biology, and Control (2018). Caister Academic Press. https://doi.org/10.21775/9781910190739
112. Jackson, R.B., Mooney, H.A., Schulze, E.-D.: A global budget for fine root biomass, surface area, and nutrient contents. Proc. Natl. Acad. Sci. USA **94**, 7362–7366 (1997). www.pnas.org
113. Jacoby, R., Peukert, M., Succurro, A., Koprivova, A., Kopriva, S.: The role of soil microorganisms in plant mineral nutrition—current knowledge and future directions. Front Plant. Sci. **8**, 1617 (2017). https://doi.org/10.3389/fpls.2017.01617
114. Jan, P., Calabrese, C., Lips, M.: Determinants of nitrogen surplus at farm level in Swiss agriculture. Nutr. Cycl. Agroecosyst. **109**, 133–148 (2017). https://doi.org/10.1007/s10705-017-9871-9

115. Jeltsch, A.: Oxygen, epigenetic signaling, and the evolution of early life. Trends Biochem. Sci. **38**, 172–176 (2013). https://doi.org/10.1016/j.tibs.2013.02.001
116. Jover, L.F., Effler, T.C., Buchan, A., Wilhelm, S.W., Weitz, J.S.: The elemental composition of virus particles: implications for marine biogeochemical cycles. Nat. Rev. Microbiol. **12**, 519–528 (2014). https://doi.org/10.1038/nrmicro3289
117. Karlen, D.L., Lal, R., Follett, R.F., Kimble, J.M., Hatfield, J.L., Miranowski, J.M., Cambardella, C.A., Manale, A., Anex, R.P., Rice, C.W.: Crop residues: the rest of the story. Environ. Sci. Technol. **43**, 8011–8015 (2006). https://doi.org/10.1021/es9011004
118. Katzourakis, A., Gifford, R.J.: Endogenous viral elements in animal genomes. PLoS Genet. **6**, e1001191 (2010). https://doi.org/10.1371/journal.pgen.1001191
119. Keen, E.C.: A century of phage research: Bacteriophages and the shaping of modern biology. BioEssays **37**, 6–9 (2015). https://doi.org/10.1002/bies.20140015230
120. Keil, D., Meyer, A., Berner, D., Poll, C., Schützenmeister, A., Piepho, H.P., VlasenkoA, Philippot L., Schloter, M., Kandeler, E., Marhan, S.: Influence of land-use intensity on the spatial distribution of N-cycling microorganisms in grassland soils. FEMS Microbiol. Ecol. **77**, 95–106 (2011). https://doi.org/10.1111/j.1574-6941.2011.01091.x (Epub 2011 Apr 15)
121. Khalifa, M.E., MacDiarmid, R.M.: A novel totivirus naturally occurring in two different fungal genera. Front. Microbiol. (2019). https://doi.org/10.3389/fmicb.2019.02318
122. Kim, K.-H., Chang, H.-W., Nam, Y.-D., Roh, S.W., Kim, M.-S., Sung, Y., Jeon, C.O., Oh, H.-M., Bae, J.-W.: Amplification of uncultured single-stranded DNA viruses from rice paddy soil Appl. Environm. Microbiol. **74**, 5975–5985 (2008). https://doi.org/10.1128/aem.01275-08
123. Kimura, M., Jia, Z.-J., Nakayama, N., Asakawa, S.: Ecology of viruses in soils: past, present and future perspectives (Review). J. Plant Nutr. Soil Sci. **4**, 1–32 (2008). https://doi.org/10.1111/j.747-0765.2007.00197.x
124. Kits, K.D., Jung, M., Vierheilig, J., Pjevac, P., Sedlacek, C.J., Liu, S., Herbold, C., Stein, L.Y., Richter, A., Wissel, H., Brüggemann, N., Wagner, M., Holger, D.: Low yield and abiotic origin of N_2O formed by the complete nitrifier *Nitrospira inopinata*. Nat. Commun. **10**, 1836 (2019). https://doi.org/10.1038/s41467-019-09790-x
125. De Klein, C.A.M., Monaghan, R.M.: The effect of farmland catchment management on nitrogen transformations and N_2O losses from pastoral systems—can we offset the effects of future intensification? Curr. Opin. Environ. Sustain. **3**, 396–406 (2011). https://doi.org/10.1016/j.cosust.2011.08.002
126. Klein, D.A.: Diversity of chytrids in biodiversity. In: Benckiser, G., Schnell, S. (eds.) Biodiv. Agricult. Product. Syst. Taylor and Francis, Boca Raton (2007)
127. Koch, H., Lücker, S., Albertsen, M., Kitzinger, K., Herbold, C., Spieck, E., Nielsen, P.H., Wagner, M., Daims, H.: Expanded metabolic versatility of ubiquitous nitrite-oxidizing bacteria from the genus Nitrospira. PNAS **112**, 11371–11376 (2015). https://doi.org/10.1073/pnas.1506533112
128. Koonin, E.V., Dolja, V.V., Krupovic, M.: Origins and evolution of viruses of eukaryotes: the ultimate modularity. Virology **479–480**, 2–25 (2015). https://doi.org/10.1016/j.virol.2015.02.039
129. Kristiansen, J., Škaloud, P.: Chrysophyta. In: Archibald, J., Simpson, A., Slamovits, C. (eds) Handbook of the Protists. Springer, Cham (2017). https://doi.org/10.1007/978-3-319-28149-0_43
130. Krupp, F., Said, N., Huang, Y.-H., Loll, B., Bürger, J., Mielke, T., Spahn, C.M.T., Wahl, M.C.: Structural basis for the action of an all-purpose transcription anti-termination factor. Mol. Cell **74**, 143–157.e5 (2019). https://doi.org/10.1016/j.molcel.2019.01.016
131. Kuzyakov, Y., Mason-Jones, K.: Viruses in soil: Nano-scale undead drivers of microbial life, biogeochemic turnover and ecosystem functions. Soil Biol. Biochem **127**, 305–317 (2018). https://doi.org/10.1016/j.soilbio.2018.09.032
132. Lander, E.S.: (International Human Genome Sequencing Consortium, 2001) Initial sequencing and analysis of the human genome. Nature **409**, 860–921 (2001). https://doi.org/10.1038/35057062

133. Lane, D.J.R., Bae, D.-H., Siafakas, A.R., Rahmanto, Y.S., Al-Akra, L., Jansson, P., Casero, Jr, R.A., Richardson, D.R.: Coupling of the polyamine and iron metabolism pathways in the regulation of proliferation: Mechanistic links to alterations in key polyamine biosynthetic and catabolic enzymes. Biochim. Biophys. Acta (BBA) Mol. Basis Dis. **1864**, 2793–2813 (2018). https://doi.org/10.1016/j.bbadis.2018.05.007
134. Lavelle, P., Decaëns, T., Aubert, M., Barot, S., Blouin, M., Bureau, F., Margerie, P., Mora, P., Rossi, J.-P.: Soil invertebrates and ecosystem services. Eur. J. Soil Biol. **42**(supplement 1), S3–S15 (2006). https://doi.org/10.1016/j.ejsobi.2006.10.002
135. Lee, B.Y., Wiringa, A.E., Bailey, R.R., Goyal, V., Tsui, B., Lewis, G.J., Muder, R.R., Harrison, L.M.: The economic effect of screening orthopedic surgery patients preoperatively for methicillin-resistant Staphylococcus aureus. Infect. Control Hosp. Epidemiol. **30**, 130–1138 (2010). https://doi.org/10.1086/657969
136. Lefeuvre, P., Martin, D.P., Elena, S.F., Shepherd, D.N., Roumagnac, P., Varsani, A.: Evolution and ecology of plant viruses. Nat. Rev. Microbiol. **17**, 632–644 (2019). https://doi.org/10.1038/s41579-019-0232-3
137. Li, Y., Wu, J., Shen, J., Liu, S., Wang, C., Chen, D., Huang, T., Zhang, J.: Soil microbial C: N ratio is a robust indicator of soil productivity for paddy fields. Sci. Rep. **6**, 35266 (2016). https://doi.org/10.1038/srep35266
138. Li, P., Lin, H., Mi, Z.: Screening of polyvalent phage-resistant *Escherichia coli* strains based on phage receptor analysis. Front. Microbiol. (2019). https://doi.org/10.3389/fmicb.2019.00850
139. Liang, X., Radosevich, M.: Commentary: A host-produced quorum-sensing autoinducer controls a phage lysis-lysogeny decision. Front. Microbiol. **10**, 1201 (2019). https://doi.org/10.3389/fmicb.2019.01201
140. Liang, X., Zhuang, J., Löffler, F.E., Zhang, Y., DeBruyn, J.M., Wilhelm, S.W., et al.: Viral and bacterial community responses to stimulated Fe(III)-bioreduction during simulated subsurface bioremediation. Environ. Microbiol. **2019**, 14566 (2019). https://doi.org/10.1111/1462-2920.14566
141. Liao, J.B.: Viruses and human cancer. Yale J. Biol. Med. **79**, 115–122 (2006) published online 2007
142. Lindow, S.E., Brandl, M.T.: Microbiology of the phyllosphere. Appl. Environ. Microbiol. **69**, 1875–1883 (2003). https://doi.org/10.1128/aem.69.4.1875-1883.2003
143. Liu, M., Deora, R., Doulatov, S.R., et al.: Reverse transcriptase-mediated tropism switching in Bordetella bacteriophage. Science **295**, 2091–2094 (2002). https://doi.org/10.1126/science.1067467
144. Liu, B., Zhang, X., Bakken, L., Snipen, L.-G., Frostegård, Å.: Rapid succession of actively transcribing denitrifier populations in agricultural soil during an anoxic spell. Front. Microbiol. (2019). https://doi.org/10.3389/fmicb.2018.03208)
145. Lovley, D.R., Walker, D.J.F.: Geobacter Protein Nanowires. Front. Microbiol. (2010). https://doi.org/10.3389/fmicb.2019.02078
146. Lozano, R., Naghavi, M., Foreman, K., Lim, S., Shibuya, K., Aboyans, V., Abraham, J., Adair, T., Aggarwal, R., Ahn, S.Y., Alvarado, M., Anderson, H.R., Anderson, L.M., Andrews, K.G., Atkinson, C., Baddour, L.M., Barker-Collo, S., Bartels, D.H., Bell, M.L., Benjamin, E.J., Bennett, D., Bhalla, K., Bikbov, B., Bin Abdulhak, A., Birbeck, G., Blyth, F., Bolliger, I., Boufous, S., Bucello, C., Burch, M., Burney, P., Carapetis, J., Chen, H., Chou, D., Chugh, S.S., Coffeng, L.E., Colan, S.D., Colquhoun, S., Colson, K.E., Condon, J., Connor, M.D., Cooper, L.T., Corriere, M., Cortinovis, M., de Vaccaro, K.C., Couser, W., Cowie, B.C., Criqui, M.H., Cross, M., Dabhadkar, K.C., Dahodwala, N., De Leo, D., Degenhardt, L., Delossantos, A., Denenberg, J., Des Jarlais, D.C., Dharmaratne, S.D., Dorsey, E.R., Driscoll, T., Duber, H., Ebel, B., Erwin, P.J., Espindola, P., Ezzati, M., Feigin, V., Flaxman, A.D., Forouzanfar, M.H., Fowkes, F.G., Franklin, R., Fransen, M., Freeman, M.K., Gabriel, S.E., Gakidou, E., Gaspari, F., Gillum, R.F., Gonzalez-Medina, D., Halasa, Y.A., Haring, D., Harrison, J.E., Havmoeller, R., Hay, R.J., Hoen, B., Hotez, P.J., Hoy, D., Jacobsen, K.H., James, S.L., Jasrasaria, R., Jayaraman, S., Johns, N., Karthikeyan, G., Kassebaum, N., Keren, A., Khoo, J.P., Knowlton, L.M., Kobusingye, O., Koranteng, A., Krishnamurthi, R., Lipnick,

M., Lipshultz, S.E., Ohno, S.L., Mabweijano, J., MacIntyre, M.F., Mallinger, L., March, L., Marks, G.B., Marks, R., Matsumori, A., Matzopoulos, R., Mayosi, B.M., McAnulty, J.H., McDermott, M.M., McGrath, J., Mensah, G.A., Merriman, T.R., Michaud, C., Miller, M., Miller, T.R., Mock, C., Mocumbi, A.O., Mokdad, A.A., Moran, A., Mulholland, K., Nair, M.N., Naldi, L., Narayan, K.M., Nasseri, K., Norman, P., O'Donnell, M., Omer, S.B., Ortblad, K., Osborne, R., Ozgediz, D., Pahari, B., Pandian, J.D., Rivero, A.P., Padilla, R.P., Perez-Ruiz, F., Perico, N., Phillips, D., Pierce, K., Pope 3rd, C.A., Porrini, E., Pourmalek, F., Raju, M., Ranganathan, D., Rehm, J.T., Rein, D.B., Remuzzi, G., Rivara, F.P., Roberts, T., De León, F.R., Rosenfeld, L.C., Rushton, L., Sacco, R.L., Salomon, J.A., Sampson, U., Sanman, E., Schwebel, D.C., Segui-Gomez, M., Shepard, D.S., Singh, D., Singleton, J., Sliwa, K., Smith, E., Steer, A., Taylor, J.A., Thomas, B., Tleyjeh, I.M., Towbin, J.A., Truelsen, T., Undurraga, E.A., Venketasubramanian, N., Vijayakumar, L., Vos, T., Wagner, G.R., Wang, M., Wang, W., Watt, K., Weinstock, M.A., Weintraub, R., Wilkinson, J.D., Woolf, A.D., Wulf, S., Yeh, P.H., Yip, P., Zabetian, A., Zheng, Z.J., Lopez, A.D., Murray, C.J., AlMazroa, M.A., Memish, Z.A.: Global and regional mortality from 235 causes of death for 20 age groups in 1990 and 2010: a systematic analysis for the Global Burden of Disease Study 2010. Lancet **380**, 2095–2128 (2012). https://doi.org/10.1016/s0140-6736(12)61728-0.ErratuminLancet2013381,628

147. Ludescher, J., Bunde, A., Schellnhuber, H.J.: Statistical significance of seasonal warming/cooling trends. PNAS **114**, E2998–E3003 (2017). https://doi.org/10.1073/pnas.1700838114
148. Madawala, H.M.S.P.: Soil-plant microbiome: a promising frontier for research. Ceylon J. Sci. **48**, 195–196 (2019). https://doi.org/10.4038/cjs.v48i3.7642
149. Marsh, M., Helenius, A.: Virus entry: open sesame. Cell **124**, 729–740 (2006). https://doi.org/10.1016/j.cell.2006.02.0074
150. Martiny, A.C.: High proportions of bacteria are culturable across major biomes. ISME J. **4**, 2125–8 (2019). https://doi.org/10.1038/s41396-019-0410-3
151. Martínez-García, M., Santos, F., Moreno-Paz, M., Parro, V., Antón, J.: Unveiling viral–host interactions within the 'microbial dark matter'. Nat. Commun. **5**, 4542 (2014). https://doi.org/10.1038/ncomms5542
152. Maruyama, F., Ueki. S.: Evolution and phylogeny of large DNA viruses, *Mimiviridae* and *Phycodnaviridae* including newly characterized *Heterosigma akashiwo* virus. Front. Microbiol. **7**, 1942 (2016). https://doi.org/10.3389/fmicb.2016.01942
153. Mateo-Sagasta, J., Raschid-Sally, L., Thebo, A.: Global wastewater and sludge production, treatment and use. In: Drechse, I.P., et al. (eds.) Wastewater (2015). Springer Science + Business Media, Dordrecht. https://doi.org/10.1007/978-94-017-9545-6_2
154. Mateo-Sagasta, J., Zadeh, S.M., Turral, H.: Water Pollution from Agriculture: A Global Review. FAO, Rome (2017). www.fao.org/land-water/overview/globalframework/global-framework, and IWMI, Colombo, www.iwmi.cgiar.org
155. McCaig, A.E., Griffiths, B.S., Prosser, J.I.: Impact of protozoan grazing on bacterial community structure in soil microcosms. Appl. Environ. Microbiol. **68**, 6094–6105 (2002). https://doi.org/10.1128/aem.68.12.6094-6105.2002
156. McLellan, E.L., Cassman, K.G., Eagle, A.J., Woodbury, P.B., Sela, S., Tonitto, C., Marjerison, R.D., van Es H.M.: The nitrogen balancing act: Tracking the environmental performance of food production. BioScience **68**, 194–203 (2018). https://doi.org/10.1093/biosci/bix164
157. Medina, G.N., Díaz-San Segundo, F., Stenfeldt, C., et al.: The different tactics of foot-and-mouth disease virus to evade innate immunity. Front. Microbiol. (2018). https://doi.org/10.3389/fmicb.2018.02644
158. Meng, Z., Li, X., Wang, H., Deng, F., Li, X., Mi, X.: Effects of converting natural grasslands into planted grasslands on ecosystem respiration: a case study in Inner Mongolia. China J. Arid Land **9**, 38–50 (2017). https://doi.org/10.1007/s40333-016-0059-y
159. Middelboe, M., Brussaard, C.P.D.: Marine viruses: key players in marine ecosystems. Viruses **9**, 302 (2017). https://doi.org/10.3390/v9100302
160. Middelboe, M., Jørgensen, N.O.G.: Viral lysis of bacteria: an important source of dissolved amino acids and cell wall compounds. J. Mar. Biol. Ass. UK, **86**, 605–661 (2006). http://dx.doi.org/0.1017/S0025315406013518

161. Millen, S., Gross, C., Donhauser, N., Mann, M.C., Péloponèse, Jr., J.-M., Thoma-Kress, A.K.: Collagen IV (COL4A1, COL4A2), a component of the viral biofilm, is induced by the HTLV-1 Oncoprotein Tax and impacts virus transmission. Front. Microbiol. **23** (2019). https://doi.org/10.3389/fmicb.2019.02439
162. Mosier, A., Parkin, T.: Gaseous emissions (CO_2, CH_4, N_2O and NO) from diverse agricultural soils. In: Benckiser, G., Schnell, S. (eds.) Biodiversity in Agricultural Production Systems, pp. 317–348. Taylor and Francis, Boca Raton pp (2007)
163. Mougari, S., Bekliz, M., Abrahao, J., et al.: Guarani Virophage, a new sputnik-like isolate from a Brazilian lake. Front. Microbiol. (2019). https://doi.org/10.3389/fmicb.2019.01003
164. Mundepi, A., Cabrera, M., Norton, J., Habteselassie, M.: Ammonia oxidizers as biological health indicators of elevated Zn and Cu in poultry litter amended soil. Water Air Soil Pollut. **230**, 239 (2019). https://doi.org/10.1007/s11270-019-4283-x
165. Nacry, P., Bouguyon, E., Gojon, A.: Nitrogen acquisition by roots: physiological and developmental mechanisms ensuring plant adaptation to a fluctuating resource. Plant Soil **370**, 1–29 (2013). https://doi.org/10.1007/s11104-013-1645-9
166. Nakatsukasa-Akune, M., Yamashita, K., Shimoda, Y., et al.: Suppression of root nodule formation by artificial expression of the TrEnodDR1 (coat protein of White clover cryptic virus 1) gene in *Lotus japonicus*. Mol. Plant Microbe Interact. **18**, 1069–1080 (2005). https://doi.org/10.1094/mpmi-18-1069
167. Naumov, A.: New relationship in carbon cycle. Low Carbon Econ. **3**, 111–114 (2012). https://doi.org/10.4236/lce.2012.323015
168. Newbold, C.J., de la Fuente, G., Belanche, A., Ramos-Morales, E., McEwan, N.R.: The role of ciliate protozoa in the rumen. Front Microbiol. **6**, 1313 (2015). https://doi.org/10.3389/fmicb.2015.01313
169. Ni, T., Zeng, Q.: Diel Infection of Cyanobacteria by Cyanophages. Front. Mar. Sci. **2**, 123 (2016). https://doi.org/10.3389/fmars.2015.0012311
170. Opitz v. Boberfeld, W. Grünlandlehre - biologische und ökologische Grundlagen, Ulmer, Germany (1994)
171. Osiander, C.N., Tafel, G.L.F., Schwab, G. (Hrsg.): First edition of Xenophon's von Athen Werke, Band 40, Von der Haushaltskunst und Hiero oder Herrscherleben. Publisher Metzler JB, Stuttgart, – In: Griechische Prosaiker in neuen Uebersetzungen (1831). https://www.booklooker.de›Bücher›Angebote›titel=Griechische+Prosai
172. De Paepe, M., Taddei, F.: Viruses' life history: towards a mechanistic basis of a trade-off between survival and reproduction among phages. PLoS Biol. **4**, e273 (2006). https://doi.org/10.1371/journal.pbio.0040193
173. Paez-Espino, D., Chen, I.A., Palaniappan, K., et al.: IMG/VR: a database of cultured and uncultured DNA viruses and retroviruses. Nucl. Acids Res. **45**, D457–D465 (2017). https://doi.org/10.1093/nar/gkw1030
174. Paez-Espino, D., Eloe-Fadrosh, E.A., Pavlopoulos, G.A., Thomas, A.D., Huntemann, M., Mikhailova, N., Rubin, E., Ivanova, N.N., Kyrpides, N.C.: Uncovering Earth's virome Nature **536**, 425–430 (2016). https://doi.org/10.1038/nature19094
175. Palatinszky, M., Herbold, C., Jehmlich, N., Pogoda, M., Han, P., von Bergen, M., Lagkouvardos, I., Karst, S.M., Galushko, A., Koch, H., Berry, D., Daims, H., Wagner, M.: Cyanate as an energy source for nitrifiers. Nature **524**, 105–108 (2015). https://doi.org/10.1038/nature14856
176. Parker, M.T.: An ecological framework of the human virome provides classification of current knowledge and identifies areas of forthcoming discovery. Yale J. Biol. Med. **89**, 339–351 PMC5045143 (2016)
177. Parton, W.J., Scurlock, J.M.O., Ojima, D.S., Schimel, D.S., Hall, D.O., Scopegram Group Members: impact of climate change on grassland production and soil carbon worldwide. Global Change Biol. **1**, 13–22 (1995). https://doi.org/10.1111/j.1365-2486.1995.tb00002.x
178. Parvathi, A., Jasna, V., Aparna, S., Ram, A.S.P., Aswathy, V.K., Balachandran, K.K., Muraleedharan, K.R., Mathew, D., Sime-Ngando, T.: High incidence of lysogeny in the oxygen minimum zones of the Arabian Sea (Southwest Coast of India). Viruses **10**, 588 (2018). https://doi.org/10.3390/v10110588

179. Pavlopoulos, G.A., Thomas, A.D., Huntemann, M., et al.: Uncovering Earth's virome. Nature **536**, 425–430 (2016). https://doi.org/10.1038/nature19094
180. Personnic, S., Domaizon, I., Sime-Ngando, T., Jacquet, Stéphan: Seasonal variations of microbial abundances and virus- versus flagellate-induced mortality of picoplankton in three peri-alpine lakes. J. Plankton Res. **31**, 1161–1177 (2009). https://doi.org/10.1093/plankt/fbp057
181. Petersen, J.M.: Microbiomes In Natura: Importance of Invertebrates in understanding the natural variety of animal-microbe interactions-host-microbe biology. mSystems **3**, e00179–1713 (2018). https://doi.org/10.1128/msystems.00179-17
182. Powel, J.M., Barros, T., Danes, M., et al.: Nitrogen use efficiencies to grow, feed, and recycle manure from the major diet components fed to dairy cows in the USA. Agri. Ecosyst. Environ. **239**, 274–282 (2017). https://doi.org/10.1016/j.agee.2017.01.023
183. Quipot, L.M.G.: ANMI fights off a vampire insect. Rice Today **11**, 16–17 (2012)
184. Ramirez, K.S., Leff, J.W., Barberán, A.., Bates, S.T., Betley, J., Crowther, T.W., Kelly, E.R., Oldfield, E.E., Shaw, E.A., Steenbock, C., Bradford, M.A., Wall, D.H., Fierer, N.: Biogeographic patterns in below-ground diversity in New York City's Central Park are similar to those observed globally. Proc. R. Soc. B **281** (2014). https://doi.org/10.1098/rspb.2014.1988
185. Raynaud, X., Nunan, N.: Spatial ecology of bacteria at the microscale in soil. Plos **9**, e87217 (2014). https://doi.org/10.1371/journal.pone.0087217
186. Reavy, B., Swanson, M.M., Cock, P.J.A., Dawson, L., Freitag, T.E., Singh, B.K., Torrance, L., Mushegian, A.R., Taliansky, M.: Distinct circular single-stranded DNA viruses exist in different soil types. Appl. Environm. Microbiol. **81**, 3934–3945 (2015). https://doi.org/10.1128/aem.03878-14
187. Renou-Wilson, F., Barry, C., Müller, C., Wilson, D.: The impacts of drainage, nutrient status and management practice on the full carbon balance of grasslands on organic soils in a maritime temperate zone. Biogeoscience **11**, 4361–4379 (2014). https://doi.org/10.5194/bg-11-4361-2014
188. Reynolds, D., Bowman, B.T., Drury, C.F., Tan, C.S., Lu, X.: Indicators of good soil physical quality: density and storage parameters. Geoderma **110**, 131–146 (2002). https://doi.org/10.1016/s0016-7061(02)00228-8
189. Reynolds, D., Drury, C.F., Tan, C.S., Fox, C.A., Yang, X.: Use of indicators of pore volume-function characteristics to quantify soil physical quality. Geoderma **152**, 252–263 (2009). https://doi.org/10.1016/j.geoderma.2009.06.009
190. Richert-Pöggeler, K.R., Franzke, K., Hipp, K., Kleespies, R.G.: Electron microscopy methods for virus diagnosis and hgh resolution analysis of viruses. Front. Microbiol. **9**, 3255 (2019). https://doi.org/10.3389/fmicb.2018.03255
191. Rodrigues, R.A.L., Andrade, A.C.S.P., Boratto, P.V.M., Trindade, G.S., Kroon, Erna G., Abrahão, J.S.: An anthropocentric view of the Virosphere-Host Relationship. Front. Microbiol. (2017). https://doi.org/10.3389/fmicb.2017.01673
192. Rohwer, F., Thurber, R.V.: Viruses manipulate the marine environment. Nature **459**, 207–212 (2009). https://doi.org/10.1038/nature08060
193. Rollins, M.C.F., Chowdhury, S., Carter, J., Golden, S.M., Miettinen, H.M., Santiago-Frangos, A., Faith, D., Lawrence, C.M., Lander, G.C., Wiedenheft, B.: Structure reveals a mechanism of CRISPR-RNA-guided nuclease recruitment and Anti-CRISPR viral mimicry. Mol. Cell **132–142**, e5 (2019). https://doi.org/10.1016/j.molcel.2019.02.001
194. Roossinck, M.J.: Move over, bacteria! Viruses make their mark as mutualistic microbial symbionts. J. Virol. **89**, 6532–6535 (2015). https://doi.org/10.1128/jvi.02974-14
195. Roossinck, M.J.: Plants, viruses and the environment: ecology and mutualism. Virology **479–480**, 271–277 (2015). https://doi.org/10.1016/j.virol.2015.03.041
196. Rosario, K., Nilsson, C., Lim, Y.W., Ruan, Y., Breitbart, M.: Metagenomic analysis of viruses in reclaimed water. Environ. Microbiol. **11**, 2806–2820 (2009). https://doi.org/10.1111/j.1462-2920.2009.01964.x
197. Roux, S., Enault, F., Robin, A., Ravet, V., Personnic, S., Theil, S., Colombet, J., Sime-Ngando, T., Debroas, D.: Assessing the diversity and specificity of two freshwater viral communities through metagenomics. Plos (2012). https://doi.org/10.1371/journal.pone.0033641

198. Roux, S., Adriaenssens, E.M., Dutilh, B.E., Koonin, E.V., Kropinski, A.M., Krupovic, M., Kuhn, J.H., Lavigne, R.J., Brister, R., Varsani, R.: Minimum information about an uncultivated Virus Genome (MIUViG). Nat. Biotech. **37**, 29–37 (2019). https://doi.org/10.1038/nbt.4306. pasteur-01977322
199. Russell, J.B., Rychlik, J.L.: Factors that alter rumen microbial ecology. Science **292**, 1119–1122 (2001). https://doi.org/10.1126/science.1058830
200. Röder, J., Dickmeis, C., Commandeur, U.: Small, smaller, nano: new applications for potato virus X in nanotechnology. Front. Plant. Sci. **10**, 158 (2019). https://doi.org/10.3389/fpls.2019.00158
201. Saleem, M., Fetzer, I., Harms, H., Chatzinotas, A.: Diversity of protists and bacteria determines predation performance and stability. ISME J. **7**, 1912–1921 (2013). https://doi.org/10.1038/ismej.2013.95)
202. Sanjuán, R.: Collective infectious units in viruses. Trends Microbiol. **25**, 402–412 (2017). https://doi.org/10.1038/ismej.2017.16
203. Scheer, C., Rowlings, D., Firrell, M., Deuter, P., Morris, S., Riches, D., Porter, I., Grace, P.: Nitrification inhibitors can increase post-harvest nitrous oxide emissions in an intensive vegetable production system. Sci. Rep. **7**, 43677 (2017). https://doi.org/10.1038/srep43677
204. Schiemann, J., Paape, M., Kellmann, J.W.: Erreger auf Wanderschaft: Transportproteine verraten pflanzenpathogenen Viren den Weg von Zelle zu Zelle. Bioforum **30**, 22–25 (2007)
205. Schils, R., Velthof, G., Mucher, S., Hazeu, G., Oenema, O., de Wit, A., Smit, A.: Current methods to estimate grassland production and biological fixation. Alterra Final Report Task 2 March 2014.docx Alterra, info.alterra@wur.nl (2013)
206. Schlüter, S., Henjes, S., Zawallich, J., Bergaust, L., Horn, M., Ippisch, O., Vogel, H.-J., Dörsch, P.: Denitrification in soil aggregate analogues-effect of aggregate size and oxygen diffusion. Front. Environ. Sci. (2018). https://doi.org/10.3389/fenvs.2018.00017)
207. Schmid, M., Speiseder, T., Dobner, T., Gonzalez, R.A.: Virus-cell interactions DNA virus replication compartments. J. Int. Virol. **88**, 1404–1420 (2014). https://doi.org/10.1128/jvi.02046-13
208. Schroeck, A.M., GaubeV, Haas E., Winiwarter, W.: Estimating nitrogen flows of agricultural soils at a landscape level—a modelling study of the Upper Enns Valley, a long-term socio-ecological research region in Austria. Sci. Total Environ. **665**, 275–289 (2019). https://doi.org/10.1016/j.scitotenv.2019.02.071
209. Schulze-Makuch, D., Wagner, D., Kounaves, S.P., Mangelsdorf, K., Devine, K.G., de Vera, J.P., Schmitt-Kopplin, P., Grossart, H.P., Parro, V., Kaupenjohann, M., Galy, A., Schneider, B., Airo, A., Frösler, J., Davila, A.F., Arens, F.L., Cáceres, L., Cornejo, F.S., Carrizo, D., Dartnell, L., DiRuggiero, J., Flury, M., Ganzert, L., Gessner, M.O., Grathwohl, P., Guan, L., Heinz, J., Hess, M., Keppler, F., Maus, K.D., McKay, C.P., Meckenstock, R.U., Montgomery, W., Oberlin, E.A., Probst, A.J., Sáenz, J.S., Sattler, T., Schirmack, J., Sephton, M.A., Schloter, M., Uhl, J., Valenzuela, B., Vestergaard, G., Wörmer, L., Zamorano, P.: Transitory microbial habitat in the hyperarid Atacama Desert. PNAS **115**, 2670–2675 (2018). https://doi.org/10.1073/pnas.1714341115)
210. Seligmann, H., Raoult, D.: Stem-loop RNA hairpins in giant viruses: invading rRNA-like repeats and a template free RNA. Front. Microbiol. **9**, 101 (2018). https://doi.org/10.3389/fmicb.2018.00101b
211. Sharma, D., Misba, L., Khan, A.U.: Antibiotics versus biofilm: an emerging battleground in microbial communities. Antimicrob. Resist. Infect. Control. **8**, 76 (2019). https://doi.org/10.1186/s13756-019-0533-3

212. Shestakov, S.V., Karbysheva, E.A.: The role of viruses in the evolution of cyanobacteria. Biol. Bull. Rev. **5**, 527–537 (2015). https://doi.org/10.1134/s2079086415060079
213. Sikorski, J.: The prokaryotic biology of soil. Soil Organ. **87**, 1–28 (2015)
214. Silpe, J.E., Bassler, B.L.: A host-produced quorum-sensing autoinducer controls a phage lysis-lysogeny decision. Cell **176**, 268–280 (2019). https://doi.org/10.1016/j.cell.2018.10.059
215. Sime-Ngando, T.: Environmental bacteriophages: viruses of microbes in aquatic ecosystems. Front. Microbiol. (2014). https://doi.org/10.3389/fmicb.2014.00355
216. Simpson, A., Slamovits, C., Archibald, J.M.: Protist diversity and eukaryote phylogeny. In: Archibald, J., Simpson, A., Slamovits, C. (eds.) Handbook of the Protists. Springer, Cham (2017). https://doi.org/10.1007/978-3-319-28149-0_45
217. Sochorec, M., Jandák, J., Raus, J., Kvasnovský, M., Hejduk, S., Knot, P.: Influence of different grassland management on water infiltration and soil physical properties Bulgarian. J. Agricult. Sci. **21**, 573–578 (2015)
218. Song, W., Xiaojuan, T., Jinsong, Z., Meng, P., Jun, L.: Autotrophic and heterotrophic components of soil respiration caused by rhizosphere priming effects in a plantation. Plant. Soil. Environ. **63**, 295–299 (2017). https://doi.org/10.17221/233/2017-PSE
219. Srinivasiah, S., Lovett, J., Polson, S., Bhavsar, S., Ghosh, D., Roy, K., Fuhrmann, J.J., Radosevich, M., Wommack, K.: Direct assessment of viral diversity in soils by random PCR amplification of polymorphic DNA. Appl. Environ. Microbiol. **79**, 5450–5457 (2013). https://doi.org/10.1128/AEM.00268-13
220. Staege, M.S., Emmer, A.: Endogenous viral elements—links between autoimmunity and cancer? Front. Microbiol. (2018). https://doi.org/10 3389/fmicb.201801371
221. Stahl, A., Vollrath, P., Samans, B., Frisch, M., Wittkop, B., Snowdon, R.J.: Effect of breeding on nitrogen use efficiency-associated traits in oilseed rape. J. Exp. Bot. **70**, 1969–1986 (2019). https://doi.org/10.1093/jxb/erz044 https://doi.org/10.1080/1343943X.2019.1617638
222. Steen, A.D., Crits-Christoph, A., Carini, P., DeAngelis, K.M., Fierer, N., Lloyd, K.G., Cameron Thrash, J.: High proportions of bacteria and archaea across most biomes remain uncultured. ISME J. (2019). https://doi.org/10.1038/s41396-019-0484-y
223. Steidinger, B.S., Crowther, T.W., Liang, J., Van Nuland, M.E., Werner, G.D.A.., Reich, P.B., Nabuurs, G., de-Miguel, S., Zhou, M., Picard, N., Herault, B., Zhao, X., Zhang, C., Routh, D., Peay, K.G.: Climatic controls of decomposition drive the global biogeography of forest-tree symbioses. Nature 569, 404–408 (2019). https://doi.org/10.1038/s41586-019-1128-0
224. Stewart, F.M., Levin, B.R.: The population biology of bacterial viruses: why be temperate. Theor. Popul. Biol. **26**, 93–117 (1984)
225. Stewart, P.S., Franklin, M.J.: Physiological heterogeneity in biofilms. Nat. Rev. Microbiol. **6**, 99–210 (2008). https://doi.org/10.1038/nrmicro1838
226. Strous, M., Pelletier, E., Mangenot, S., Rattei, T., Lehner, A., Taylor, M., Horn, M., Daims, H., Bartol-Mavel, D., Wincker, P., Barbe, V., Fonknechten, N., Vallenet, D., Segurens, B., Schenowitz-Truong, C., Médigue, C., Collingro, A., Snel, B., Dutilh, B.E., Op den Camp, H.J.M., van der Drift, C., Cirpus, I., van de Pas-Schoonen, K.T., Harhangi, H.R., van Niftrik, L., Schmid, M., Keltjens, J., van de Vossenberg, J., Kartal, B., Meier, H., Dj, Frishman, Huynen, M.A., Mewes, H.-W., Weissenbach, J., Jetten, M.S.M., Wagner, M., Le Paslier, D.: Deciphering the evolution and metabolism of an anammox bacterium from a community genome. Nature **440**, 790–794 (2006). https://doi.org/10.1038/nature04647
227. Tebbe, C.C., Schloter, M.: Discerning the diversity of soil prokaryotes (bacteriaand archaea) and their impact on agriculture. In: Benckiser, G., Schnell, S. (eds.) Biodiversity in Agricultural Production Systems, pp. 81–100. Taylor and Francis, Boca Raton pp (2007)
228. Tennant, P., Fermin, G., Foster, J.E. (eds.): Viruses: Molecular Biology, Host Interactions, and Applications to Biotechnology, p. 224. Academic Press, London (2018)
229. Thomas, R., Berdjeb, L., Sime-Ngando, T., Jacquet, S.: Viral abundance, production, decay rates and life strategies (lysogeny versus lysis) in Lake Bourget (France). Envron. Microbiol. **13**, 616–630 (2011). https://doi.org/10.1111/j.1462-2920.2010.02364.x
230. Thompson, L.R., Zeng, Q., Kelly, L., et al.: Phage auxiliary metabolic genes and the redirection of cyanobacterial host carbon metabolism. PNAS **108**, 757–764 (2011). https://doi.org/10.1073/pnas.110216410810

231. Thomson, B.C., Ostle, N.J., McNamara, N.P., Oakley, S., Whiteley, A.S., Bailey, M.J., Griffiths, R.I.: Plant soil interactions alter carbon cycling in an upland grassland soil. Front. Microbiol. (2013). https://doi.org/10.3389/fmicb.2013.00253
232. Thézé, J., Bézier, A., Periquet, G., Drezen, J.-M., Herniou, E.A.: Paleozoic origin of insect large dsDNA viruses. PNAS **108**, 15931–15935 (2011). https://doi.org/10.1073/pnas.1105580108
233. Torsvik, V., Daae, F.L., Sandaa, R.-A., Øvreås, L.: Novel techniques for analysing microbial diversity in natural and perturbed environments. J. Biotechnol. **64**, 53–62 (1998). https://doi.org/10.1016/s0168-1656(98)00103-5
234. Tsujimoto, Y., Rakotoson, T., Tanaka, A., Saito, K.: Challenges and opportunities for improving N use efficiency for rice production in sub-Saharan Africa. Plant Prod. Sci. **22**, 413–427 (2019). https://doi.org/10.1080/1343943x.2019.1617638
235. Turkington, C.J.R., Morozov, A., Clokie, M.R.J., Bayliss, C.D.: phage-resistant phase-variant subpopulations mediate herd immunity against bacteriophage invasion of bacterial metapopulations. Fron Microbiol. (2019). https://doi.org/10.3389/micb.2019.01473
236. Villamar, C.-A., Vera-Puerto, I., Rivera, D., De la Hoz, F.: Reuse and recycling of livestock and municipal wastewater in Chilean agriculture: a preliminary assessment. Water **10**, 817 (2018). https://doi.org/10.3390/w10060817
237. Vincent, S.J., Coutts, B.A., Jones, R.A.C.: Effects of introduced and indigenous viruses on native plants: exploring their disease causing potential at the agro-ecological interface. PLoS ONE **9**, e91224 (2014). https://doi.org/10.1371/journal.pone.0091224
238. Vinoth Kumar, R.: Plant antiviral immunity against geminiviruses and viral counter-defense for survival. Front. Microbiol. (2019). https://doi.org/10.3389/fmicb.2019.01460
239. Vogel, H.J., Babel, U.: Soil space diversity and its dynamics: qualitative and quantitative considerations. In: Benckiser, G., Schnell, S. (eds.) Biodiversity in agricultural production systems. Taylor & Francis, Boca Raton (2007)
240. Volle, R., Razafindratsimandresy, R., Joffret, M.-L., Bessaud, M., Rabemanantsoa, S., Andriamamonjy, S., Raharinantoanina, J., Blondel, B., Heraud, J.-M., Bailly, J.-L., Francis, Delpeyroux: High permissiveness for genetic exchanges between enteroviruses of species A, including enterovirus 71, favours evolution through intertypic recombination in Madagascar. J Virol. (2019). https://doi.org/10.1128/jvi.01667-18
241. Vrede, K., Heldal, M., Norland, S., Bratbak, G.: Elemental composition (C, N, P) and cell volume of exponentially growing and nutrient-limited bacterioplankton. Appl. Environ. Microbiol. **68**, 965–2971 (2002). https://doi.org/10.1128/aem.68.6.2965-2971.2002
242. Wagner, M.: Single-cell ecophysiology of microbes as revealed by Raman microspectroscopy or secondary ion mass spectrometry imaging. Ann. Rev. Microbiol. **63**: 411–429 (2009). https://doi.org/10.1146/annurev.micro.091208.073233
243. Wang, P., Xu, J., Wang, Y., Cao, X.: An interferon-independent lncRNA promotes viral replication by modulating cellular metabolism. Science **358**, 1051–1055 (2017). https://doi.org/10.1126/science.aao0409
244. Weinbauer, W.G.: Ecology of prokaryotic viruses. FEMS Microbiol. Rev. **28**, 127–181 (2004). https://doi.org/10.1016/j.femsre.2003.08.001
245. Weitz, J.S., Dushoff, J.: Alternative stable states in host-phage dynamics. Theor. Ecol **1**, 13–19 (2008). https://doi.org/10.1007/s12080-007-0001-1
246. Weitz, J.S., Wilhelm Steven, W.: Ocean viruses and their effects on microbial communities and biogeochemical cycles. F1000 Biol. Rep. **4**, 17 (2012). https://doi.org/10.3410/b4-17
247. Whipps, J.M., Lewis, K., Cooke, R.C.: Mycoparasitism and plant disease control, pp. 161–187. In: Burge, M.N. (ed.) Fungi and Biological Control Systems. Manchester University Press, Manchester (1988)
248. Whitaker, B.K., Rua, M.A., Mitchel, C.E.: Viral pathogen production in a wild grass host driven by host growth and soil nitrogen. New Phytol. **207**, 768 (2015). https://doi.org/10.1111/nph.13369
249. White, R.R.: Increasing energy and protein use efficiency improves opportunities to decrease land use, water use, and greenhouse gas emissions from dairy production. Agric. Syst. **146**, 20–29 (2016). https://doi.org/10.1016/j.agsy.2016.03.013

250. Whitfield, A.E., Falk, B.W., Rotenberg, D.: Insect vector-mediated transmission of plant viruses. Virology **479–480**, 278–289 (2015). https://doi.org/10.1016/j.virol.2015.03.026
251. Wilhelm, S.W., Suttle, C.A.: Viruses and nutrient cycles in the sea. Bioscience **49**, 781–788 (1999)
252. Williamson, K.E., Fuhrmann, J.J., Wommack, K.E., Radosevich, M.: Viruses in soil ecosystems: an unknown quantity within an unexplored territory. Annu. Rev. Virol. **29**, 201–219 (2017). https://doi.org/10.1146/annurev-virology-101416-041639
253. Williamson, K.E.: Soil phage ecology: abundance, distribution, and interactions with bacterial hosts. In: Wiltzany, G. (ed.) Bio-Communication in Soil Microorganisms. Soil Biology, vol. 23, pp 113–136. Springer, Berlin (2011)
254. Wilpiszeski, R.L., Aufrecht, J.A., Retterer, S.T., Sullivan, M.B., Graham, D.E., Pierce, E.M., Zablocki, O.D., Palumbo, A.V., Elias, D.A.: Soil aggregate microbial communities: towards understanding microbiome interactions at biologically relevant scales. Appl. Environ. Microbiol. **85**, e00324–19 (2019). https://doi.org/10.1128/aem.00324-19
255. Wu, G.A., Jun, S.-R., Sims, G.E., Kim, S.-H.: Whole-proteome phylogeny of large dsDNA virus families by an alignment-free method. PNAS **106**, 12826–2831 (2009). https://doi.org/10.1073/pnas.0905115106
256. Xiao, X., Melillo, M., Kicklighter, D.W., McGuire, A.D., Prinn, R.G., Wang, C., Stone, P.H., Sokolov, A.: Transient climate change and net ecosystem production of the terrestrial biosphere. Global Biochem. Cycl. **12**, 345–360 (1998). https://doi.org/10.1029/98GB01035
257. Xu, P., Liu Y, Graham, R.I., et al.: Densovirus is a mutualistic symbiont of a global crop pest (*Helicoverpa armigera*) and protects against a baculovirus and Bt biopesticide. PLoS Pathog. **10**, e1004490 (2014). https://doi.org/10.1371/journal.ppat.1004490
258. Yaeger, R.G.: Protozoa: structure, classification, growth, and development. In: Baron, S. (ed.) Medical Microbiology, 4th edn. (Chap. 77) (1996). University of Texas Medical Branch at Galveston, Galveston (TX). Available from: https://www.ncbi.nlm.nih.gov/books/NBK8325/
259. Yang, X., Chen, B., Zhang, T., et al.: Geographic distribution and genetic diversity of rice stripe mosaic virus in Southern China. Front. Microbiol. (2018). https://doi.org/10.3389/fmicb.2018.03068
260. Ye, J., Perez, P.G., Zhang, R., Nielsen, S., Huang, D., Thomas, T.: Effects of different C/N ratios on bacterial compositions and processes in an organically managed soil. Biol. Fertil. Soils **54**, 137–147 (2018). https://doi.org/10.1007/s00374-017-1246-5
261. Yeates, G.W.: Diversity of nematodes. In: Benckiser, G., Schnell, S. (eds.) Biodiversity in Agricultural Production Systems, pp. 215–235. Taylor and Francis, Boca Raton (2007)
262. Young, R.F., White, R.L.: Lysis of the host by bacteriophage. In: Mahy, B.J.W., van Regenmortel, M.H.V.: (eds.) Encyclopedia of Virology, 3rd edn. Elsevier Ltd., Oxford, pp. 144–158 (2008)
263. Yu, H., Ling, N., Wang, T., Zhu, C., Wang, Y., Wang, S., Gao, Q.: Responses of soil biological traits and bacterial communities to nitrogen fertilization mediate maize yields across three soil types. Soil Tillage Res. **185**, 61–69 (2019). https://doi.org/10.1016/j.still.2018.08.017
264. Zhang, X., Hu, H.-Y., Men, Y., Yang, Y., Christoffersen, K.S.: Feeding characteristics of a golden alga (*Poterioochromonas* sp.) grazing on toxic cyanobacterium *Microcystis aeruginosa*. Water Res. **43**, 2953–2960 (2009). https://doi.org/10.1016/j.watres.2009.04.003
265. Zhang, Y.Z., Shi, M., Holmes, E.C.: Using metagenomics to characterize an expanding virosphere. Cell **172**, 1168–1172 (2018). https://doi.org/10.1016/j.cell.2018.02.043
266. Zhang, J., Elser, J.J.: Carbon:Nitrogen:Phosphorus stoichiometry in fungi: a meta-analysis. Front. Microbiol. **8**, 1281 (2017). https://doi.org/10.3389/fmicb.2017.01281
267. Zhivich, A.: Fighting bacterial resistance: approaches, challenges, and opportunities in the search for new antibiotics. In: Part 1. Antibiotics used in clinical practice: mechanisms of action and the development of bacterial resistance MIR J 4 (2017). https://doi.org/10.18527/2500-2236-2017-4-1-31-51

268. Zhou, L., Zhao, B., Zhang, L., Wang, S., Dong, D., Lv, H., Shang, P.: Alterations in cellular iron metabolism provide more therapeutic opportunities for cancer. Int. J. Mol. Sci. **19**, 1545 (2018). https://doi.org/10.3390/ijms19051545
269. Zhu, D., Palatinszky, M., Schintlmeister, A., Schmid, M.C., Hanson, B.T., Shterzer, N., Mizrahi, I., Rauch, I., Decker, T., Bocklitz, T., Popp, J., Gibson, C.M., Fowler, P.W., Huang, W.E., Wagner, M.: Tracking heavy water (D_2O) incorporation for identifying and sorting active microbial l cells. PNAS **112**, E194-E203 (2015). https://doi.org/10.1073/pnas.1420406112
270. Zunke, U., Perry, R.N.: Nematodes: harmful and beneficial organisms. In: Benckiser, G. (ed.) Fauna in Soil Ecosystems—Recycling Processes, Nutrient Fluxes, and Agricultural Production, pp. 85–133. Marcel Dekker, New York, USA) (1997)

N Fertilization Dependent Bacterial and Archaeal Changes in Paddy Soil

Sakthivel Ambreetha, Kalyanasundaram Geetha Thanuja, Subburamu Karthikeyan, and Dananjeyan Balachandar

Abstract Paddy ecosystem naturally serves as a micro-habitat for diversified soil microbiome and its oxic-anoxic interface is the hub of multiple bio-geochemical cycles. However, in the Anthropocene era, paddy fields are not exempted from vigorous human interventions. On one hand the cropping of short duration high yielding varieties requiring high nutrient inputs and the other specifically, the heavy dosage of nitrogen inputs often extremely higher than the recommended dose. Furthermore, the inefficient agricultural practices lead to loss of applied N through volatilization, leaching, and low nitrogen use efficiency. In all the cases, imbalance in N level instantly alters the diversity of N-dependent bacterial and archaeal populations. Asia contributes to 90% of global rice production and many of its agricultural lands are constantly occupied with year-round paddy crops for decades. In such areas, prolonged imbalance in N might have a serious impact on soil microbiome subsequently collapsing the natural N transformations in the paddy ecosystem. The abundance of N transforming microbes such as nitrogen-fixing bacteria, ammonia-oxidizing bacteria/archaea, nitrite oxidizers, nitrate oxidizers, denitrifying bacteria, and anaerobic ammonia oxidizers are the key players in sustainable paddy cultivation. Planning a balanced N fertilization certainly helps in proper maintenance of soil health to ensure persistent crop production and sustainability. Hence, a detailed understanding of the adverse impact of higher and sub-optimal dosage of N fertilizers on paddy microbiome is critical. This chapter details the importance of microbial-mediated N transformations in paddy ecosystem followed by the impact of imbalanced N fertilization over soil microbial abundance and the soil health thereof.

Keywords Paddy · Soil dynamics · N fertilization · Core microbiome · Soil health

---

S. Ambreetha · K. G. Thanuja · S. Karthikeyan · D. Balachandar (✉)
Department of Agricultural Microbiology, AC&RI, Tamil Nadu Agricultural University, Coimbatore 641003, India
e-mail: dbalu@tnau.ac.in

S. Karthikeyan
Department of Renewable Energy Engineering, AEC & RI, Tamil Nadu Agricultural University, Coimbatore 641003, India

G. Benckiser (ed.), *Soil and Recycling Management in the Anthropocene Era*, Environmental Science and Engineering,
https://doi.org/10.1007/978-3-030-51886-8_3

1 Paddy as Microhabitat

Rice (*Oryza sativa* L.) serves major population in the world as a staple food crop cultivated with the area of 160 million ha which meets dietary requirement of nearly half the globe [24]. However, the global rice production falls under both irrigated and rainfed conditions, they are predominantly cultivated in flooded systems. Rice still stands as a dynamic model for number of ecological and environmental studies besides their sustainability and nutrient recycling capacity [8] and regarded globally as largest man-made wetland system. Prolonged flooded condition for the cultivation of crop is a fore runner for organic matter accumulation and carbon management. Growing population puts pressure on land to multiply the yield to feed them which sets step for numerous management practices which paved way for intensive fertilization. Driven towards the desire for greater yields, imbalanced and over fertilization posed series of environmental concerns [49] like decreased nutrient use efficiency, adverse effect on soil microbial biomass and community composition [35] etc., Input nutrient utilization and recycling by active microbes in the soil serve as a reason making the system productive. Redox process under flooded system governs nutrient availability, their cycling and transformations, and ecological functions which is an add-on reason for holding unique functions. On comparison with upland crop, lowland paddy systems are conquered by distinct biotransformation of carbon and nitrogen mediated by microbial communities [34]. Over the last decades, there is a substantial increase in rice yield with the use of various N fertilizers [68]. On the other hand, microbial soil processes affect N dynamics and therefore it is important to study their interrelation with microbial community. Microbial flora which embodies healthy soil is affected by various management practices in which fertilization plays pivotal role. In a study in long-term permanent manurial experimental plot, the fractions of soil organic carbon, microbial biomass carbon, culturable microbial counts and dehydrogenase and β-glucosidase activities increased significantly in green manure amended soil. Further higher nitrifier population was evident in inorganic or integrated organic and inorganic fertilizer [73].

1.1 Soil Characteristics

Rice cultivation under flooded condition exhibits distinct physico-chemical properties and encounters diversified microflora responsible for numerous biochemical activities. They serve as an excellent model for acquiring insights on soil ecological processes [103]. The biological changes endowed with them paves way for typical succession of chemical transformations. The soil under submergence creates condition reducing the oxygen supply into the soil whose diffusivity in water is 10,000 times lower than in air [43]. Meanwhile, organic matter and photosynthates accumulation [115] contribute for reduction process in soil and oxidation of organic matter causing low oxidation—reduction potential. The intermittent change of redox potential (Eh) is a unique character, which vary from −200 to −300 mV to + 500 to + 700 mV. Reduced environment with an Eh of—250 mV is prevalent after flooding when the soil pores are filled with water and free oxygen replacing air and soil environments. Alternate wetting and drying process encourages change in pH and are influenced by type of cations, water contents and concentration of neutral salts. The variation in both pH and Eh approves changes in redox sensitive components like iron, manganese, sulfur, carbon and nitrogen [78]. The reaction curtails oxygen supply furthermore and results in death of aerobic organisms and proliferation of facultative anaerobes. They are in turn proceeded by series of reduction process and the reducing elements in the order of O_2, NO_3, MnO_2 and Fe $(OH)_3$-Fe^{2+} resulting in reduced counterparts namely Fe(II), Mn(II),NH_4 along with an array of unstable organic components [72]. This also results in domination of facultative microbes followed by anaerobic microorganisms consequently using oxidized soil substrate as electron acceptors.

1.2 Zones of Wetland

The assorted soil-water lowland system comprises of three zones viz., surface (oxic), anoxic and rhizosphere which harbor indigenous microbial community with its own unique biological activities respectively [69]. The zone surrounding the plant roots gets accustomed to submerged soils and develops process ensuring oxygen and other gas exchange mainly through aerenchyma tissues. The porous aerenchyma enables root respiration where most of the oxygen is consumed and escaped oxygen oxidizes the reduced soil components surrounding them making thin aerobic zone characterized by high redox potential [95]. The oxic layer can further be differentiated into floodwater and oxidized surface layer which are oxidized and partially oxidized respectively. Oxidized surface layer is the water column superimposed on the anaerobic layer marked by dissolved oxygen and ranges from few mm to one cm in thickness depending upon the O_2 utilization rate [32]. They are characterized with higher redox potential encompassing aerobic microbial reactions like nitrification, ammonification, and immobilization. They are also prone to pH fluctuations due growth of algae which ensures removal of CO_2. The anaerobic (anoxic) layer is characterized by low redox potential, holds few cm thickness and are conducive accumulation of ammonia [50]. Waterlogged cultivation of paddy supports anoxia in most of soil compartments rendering a hotspot for methanogenenic microorganisms. In addition, various N transformation reactions including N_2 fixation, N mineralization and immobilization, denitrification and dissimilatory nitrate reduction occurs. Thus, anaerobic metabolism heads the role and their products formed marks distinct from the arable soils.

2 Biogeochemistry of N in Paddy Soil

Nitrogen being associated to ecosystem stability and primary productivity, the knowledge on transformation mediated by microbes is significant in N cycling [44]. Especially the rate of N deposition on the terrestrial ecosystem is expected to project by a factor of 2.5 in forth coming century [117]. They in turn put forth various consequences including change in microbial community structure, activities and ultimately altering ecological functions. In paddy ecosystem, it serves as an important element during early vegetative and panicle initiation stage and regarded as limiting nutrient for productivity [92]. They are unique from other nutrients, since their role in formation of chlorophyll, proteins and nucleic acids is most significant. The N input for the crop includes fertilizers, source of irrigation, precipitation and biological N_2 fixation, while the output includes crop removal, volatilization, leaching/run off, denitrification. Hence their management is a mandate to procure high yield and knowledge on available N from soil organic matter and added fertilizer would assist in N management technology to ensure sufficient food grain production [10]. Intensive paddy cultivation recruits high N application which goes nearly 500 kg ha^{-1} per crop [14]. Nitrogen transformation under flooded condition influences microbial structure and function, turnover rate, availability and losses of N which differs prominently from drained or aerated soil. N cycling in paddy soil is affected mainly by depletion of oxygen and includes ammonification, N immobilization, nitrification, denitrification, anaerobic ammonium oxidation (anammox) and nitrogen fixation [42]. However, these processes were dug in detail, the significant shifts on N cycling communities and activities stands on the fore to be explored.

2.1 N Cycling

Nitrogen is the crucial nutrient for all live forms, however most organisms bank on bioavailable forms [57]. The network of nitrogen transforming microbes strongly persists in paddy ecosystem and had been affected by input of intensive N fertilizers. It is noteworthy to explore versatility of microbes involved in N transformation and the biochemical pathway.

2.1.1 Ammonification

Organic nitrogen supplied to the soil is mineralized (ammonification / N mineralization) or broken down by anaerobic decomposition aided by heterotrophic bacteria and fungi [11]. A key process which determines the availability of N and wetland productivity occurs in oxidized surface layer, reduced soil and also in rhizosphere, where ammonia volatilization is followed in flooded water [50]. The rate of ammonification is lower because of deprived O_2 status under anoxic zones. Savant and De Datta [82] elucidated ammonification which comprises of oxidative and reductive deamination of amino acids operative under oxidized and reduced soil layers, respectively. This reaction secretes part of ammonia into soil solution while the remaining into adsorbed phase where there is a dynamic equilibrium occurs between them. Ponnamperuma [72] have elucidated the portion ammonia in the solution phase depends on organic matter and texture of the soil. Their rate is comparatively higher in the former and even influenced by certain parameters like pH, temperature, C: N ratio, soil moisture and organic matter status. Anaerobic condition is marked by accumulation of ammonium, nitrate instability which ultimately leads to lowered N requirement for decomposition of organic matter. However, the degree of mineralization is rapid under warm temperature,the soil water content is insensitive. Under elevated CO_2 and temperature, the presence of higher C substrates with labile N pools in rhizosphere positively influenced the N mineralization [7]. Apart from this, the labile N pools including microbial biomass N, ammoniacal N and nitrate N were significantly higher under elevated CO_2 and temperature than that of ambient conditions. Ammonium—N accumulates in reduced conditions notably on the soil with high organic load, while on the other hand in soil with lower organic matter ammonium is rapidly converted to nitrite and further to nitrate. Until when rice roots uptake or losses occur, NH_4^+-N remains stable under reduced conditions and retains as cations and converted into cations upon the exposure of oxygen. Ammonia volatilization is considered as major N loss and receives considerable attention for their mitigation. He et al. [39] have observed high flux of ammonia volatilization in the absence of water table. They are regulated by properties of flooded water such as pH and temperature. These results were similar to the study conducted by Hayashi et al. [38], suggesting their loss could be handled with appropriate management practices.

2.1.2 N Immobilization

The balance between nitrogen immobilization and mineralization is the driving factor for the availability of N required for plant uptake [23]. The equilibrium between them is influenced by various factors such as oxygen status, temperature, moisture and organic material and C:N ratio. It is one of the most critical steps influencing the long term fertility of the soil. Reduced fertilizer use efficiency upon their application sets step for N losses with greater degree of immobilization. The higher extent of immobilization paves way for low N fertilizer use efficiency. Immobilization takes route viz., microbial biomass, amino acids and amino sugars [67]. They are further remineralized into fractions of soil organic matter. Anaerobic conditions prevailing under paddy cultivation immobilize greater N from the biomass. The rate of N immobilization further increases upon rice straw incorporation suggesting the importance the crop residue and irrigation management practices in N immobilization [79]. In depth acquaintance on biological control on various route of immobilization would facilitate better utilization of N from crop residues and fertilizers.

2.1.3 Nitrification

The aerenchymatous tissue present in rice crop allows diffusion of oxygen into rice roots and soil which favors nitrification in aerobic niche. Thus the rice roots are exposed to mixture of NH_4^+–NO_3^-. The anoxic environment in paddy encourages microbial reduction of available NO_3^- making their availability low. The unceasing supply of NO_3^- is warranted by oxic conditions prevailing in rhizosphere by their oxygen supply. This chemolithotrophic nitrification involves two steps involving ammonia oxidation to nitrite, and nitrite oxidation to nitrate [17] which is performed by ammonia oxidizing bacteria or archaea (AOB/AOA) [55] and nitrite oxidizing bacteria (NOB) [21] respectively. The activity shoots up upon the addition of N fertilizer and AOB population like *Nitrosomonas, Nitrosococcus* which engrosses in biological nitrification than AOA, however rice rhizosphere encompasses both AOA and AOB. AOB communities are in turn decided by several factors including soil temperature, pH, availability of ammonium and oxygen. With increase in N fertilization, the relative abundance of *Nitrosomonas* was found to be higher than that of *Nitrospira.* The reason could be due to increased pore water NH_4^+- N concentration upon N fertilization. Nitrite oxidation is carried out by genera belonging to *Nitrobacter* and *Nitrococcus*. Recent nitrification research revealed certain common microorganisms taking up role of both the process belonging to the genera *Nitrospira* and termed as commamox [20, 97]. Commamox are contributors of nitrification in paddy soil even at low ammonia niche than AOA and AOB characterized by higher *amoA* gene abundance [60]. Anoxic condition prevailing enables the reduction of NO_3^- thereby limiting their availability. However the oxygen leakage from the rice roots enables the continuous supply of NO_3^-.

2.1.4 Denitrification

The major nitrogen removal mechanism characterized by respiratory transformation of inorganic N (oxides of nitrogen) to gaseous forms such as nitric oxide, nitrous oxide and N_2 [74, 89]. This process occurs just below the oxidized zone where nitrification occurs and accounts for more than 40% of N fertilizer loss [120, 122]. This is considered as major contributor for nitrogen loss from applied nitrogenous fertilizers and source for greenhouse gases. When the redox potential decreases after 10–14 days of water logging, denitrifier activity becomes higher. Nitrate ion conversion to di-nitrogen gas has been studied in detail by [72], and elucidated their extreme instability. This gives rise to potential denitrification activity in paddy ecosystem. This reaction involves the reduction of nitrite to nitric oxide catalyzed by key enzyme nitrite reductase [15]. In general, N fertilization coupled with flooded condition stimulates denitrifying bacteria, which denitrifies the nitrate in anaerobic soil. Facultative and heterotrophic anaerobes use NO_3^- as terminal acceptor involving them redox reaction between organic carbon and oxides of N. Organic carbon serves as electron donor for the process and their availability plays a prime role. It was further supported by [108] that nitrification, provides substrate for denitrification and creates competition between nitrifiers and rice plants for NH_4^+. Coupled reaction of nitrification and denitrification is supported by oxic/anoxic interface which give rise NO_3^- as intermediate product. Nitrification activity upon the addition of N fertilizer has a partitioning activity of NO_3^- between denitrification and dissimilatory nitrate reduction to ammonium (DNRA). DNRA is one among the important NO_3^- reduction processes, which enables the conversion of loss prone NO_3^- into retainable NH^{4+} [111]. The rate of nitrification is one of the major factor governing the nitrogen use efficiency and N losses through denitrification. Nitrification of NH_4^+ into NO_3^-, indeed serves as a substrate for denitrification, when plant uptake of NH_4^+ is reduced [108].

The enzymes involved in the process include nitrate reductase, nitrite reductase, nitric oxide reductase and nitrous oxide reductase [15]. Among them, nitrite reductase is considered as key enzyme, *nir*K and *nir*S as functional genes, which is widely used as a molecular marker for denitrifiers. *nir*K and *nir*S are structurally distinct and functionally equivalent, copper and cytochrome cd1 containing nitrite reductase respectively. Braker et al. and Yoshida et al. [9, 109] have elucidated the diversity and quantity *nirS* and *nirK*. *NirS* diversity was found more in paddy system belonging to Burkholderiales, Rhodocyclales and Rhodobacteriales, while *nirK* was related to rhizobiales.

2.1.5 Anaerobic Ammonium Oxidation (anammox)

The anaerobic oxidation of ammonium along with nitrite is a significant transformation process affecting the global N budget. Anoxic condition usually harbors enormous NH_4^+, while the supply NO_2^- is the limited. The condition is driven by two ways either by oxidation of NH_4^+ to NO_2^- by AOB or from NO_3^- reduction by denitrifying bacteria. Welte et al. [104] have reported nitrite dependent and nitrite independent anaerobic methane oxidation in paddy fields, where the activity was higher in top 20 cm of the soil. Similar results were found by [118] with higher anommox activity in top layers. In contrast, their activity was even found in deeper layers of continuously flooded systems. Zhu et al. and Ding et al. [121, 25] have reported NH_3 oxidation pathway coupled with ferric minerals in paddy soil for the first time. It is commonly called as feammox, involving dissimilatory Fe reducing microbes [85] whose activity is significantly higher in rhizosphere [93]. They recruit different microbes yielding NO_2^-, NO_3^- and N_2 as terminal products. Denitrification and annamox reaction responsible for NO_3- reduction contribute to significant losses of N, which are equilibrated by DNRA. DNRA occurs when nitrate in the soil is limiting. They in turn enhances N retention transforming NO_3^- into soil retainable NH_4^+ [111].

2.1.6 Nitrogen Fixation

The biological conversion of atmospheric nitrogen in air to ammonia in soil is accomplished by archaea and bacteria with the help of nitrogenase encoded genes. This reduction process in flooded ecosystem is supported by low oxidation and reduction potentials and also by the organic matter. Biological nitrogen fixation (BNF) plays pivotal role in flooded paddy system in maintaining the long term soil productivity [58]. Diversity of diazotrophs in the rhizosphere is enhanced by lower NH^{4+} –N concentration. The process plays pivotal role in supplying N to the plant, where no chemical fertilizers are used [76]. Since nitrogen fixing photosynthetic bacteria demands anaerobic condition, reduced region supports the process than the oxidized zone. The rate of nitrogen fixation ranges between 0.03 and 46.2 g N m^{-2} y^{-1} within the redox range from −200 to −260 mV [10]. The complex and dynamic ecosystem of paddy encompasses variety of N_2 fixers including rice associative and free living heterotrophs, free living and symbiotic photosynthetic cyanobacteria. Metagenome analysis of rhizosphere and phyllosphere evidenced the transcripts of *nifH*, *nifD* and *nifK*. Diazotrophic community in paddy is unstable since various environmental factors and fertilizer application influences their function [91].

2.2 *Fate of N in Paddy Ecosystem*

N fertilizer applied to the soil is either partly absorbed by the plant or fixed by the soil, while the rest gets lost in different routes [5]. For producing one ton of rough rice, the crop removes 16–17 kg of N. The amount of N added to the soil and the quantity used by the plant is mostly unbalanced and interrupted by number of factors. High quantity of N applied into the soil, is lost and favored by layers of the soil for aerobic and anaerobic N metabolism. In the study of N leaching effect, [65] have hypothesized that more loss of NO_3^- than NH_4^+, which is due to NH_4^+ adsorbance into the soil particles. Studies on ^{15}N urea for the detection of recovery and residual rate accounted only 16–25%, while non-specified N percentage rate was higher [71]. These results suggest that continuous flooding intensifies N loss in various ways. Besides N loss transformation process, there occurs several temporal variability as a function of several management practices mainly irrigation and crop residues [18].

3 Microbiome of Paddy

Microbiome refers to the totality of microbial communities present in a particular habitat. The microbial communities remain diversified within a habitat based on various external factors including human intervention. Nevertheless, certain microbiome will be universally predominant among all the microbial assemblages of a particular ecosystem. The microbial community which flourish in a specific habitat irrespective of space and time are designated as core microbiome [83]. In crisp, it includes the population essential for the functioning of that particular ecosystem. More evidently, scientists have put forth the possibility of positive correlation between microbial community structure and the functionality of an ecosystem [22]. With the foresaid concept, exploring the core microbiome may ultimately pave way for better operation of paddy ecosystem wherein the role of microbes remain inevitable. Undermining the paddy core microbiome is counted as need of the hour as rice is the staple food for most of Asian countries. Despite the significance, delineating core microbiome is way more lagging due to the daunting efforts required for thorough understanding of microbial behaviour across distantly located paddy habitats.

However, a recent continental survey across China has unleased the biogeographic pattern of microbial distribution in paddy ecosystem. The study has disentangled the myriad roles of α- and β-diversity of core microbiome in multi-nutrient transformations in paddy field [47]. Paddy ecosystem serves as a hub for N and C biogeochemical cycles mediated by the associated microbiome. Evidently, nitrite-dependent methane oxidizing anaerobe *Candidatus Methylomirabilis oxyfera* has been reported to bridge both C and N cycling in paddy fields [119]. Paddy ecosystem is generally occupied by diversified group of ammonia oxidizing archaea and bacteria, nitrifiers, denitrifiers, methanogens, methanotrophs etc. actively involved in C and N cycling. These active microbiota associated with paddy undergo spatial variations based on different layers of soil and distance from the rhizospheric zone of the crop [63]. Differential assemblage of paddy microbiome has documented the occurrence of *Xanthomonadales*, *Myxococcales*, and *Methylococcales* in oxic surface layer of soil and predominance of methanogenic archaea and other bacterial anaerobes in anoxic bulk soil [53]. In yet another outlook, an interdisciplinary approach to delineate core microbiota has demarcated the evidence for the ability of core microbiome to recruit beneficial native flora while suppressing the plant pathogens [94]. Aforesaid evidence justifies the need of core microbiome for proper nutrient cycling in paddy ecosystem along with their significant role in plant health and sustainability.

3.1 K and R Strategic Groups

Microbial community in any ecosystem can be physiologically classified into K and R strategic groups. K strategic microbes, nutritionally called oligotrophs are slow growers whose population does not rely on nutritional fluctuations in soil. In contrary, R strategic or copiotrophic groups have high fecundity and get altered based on external inputs [30]. For example, N fertilization in paddy field increased the abundance of R strategists belonging to Proteobacteria and Bacteroidetes. Additionally, it exerted an indirect negative impact on K strategists such as Acidobacteria as the shoot up of R strategists outcompeted other slow growing bacteria in the soil [31]. While delineating the core microbiome of paddy, metagenomic approaches would better help to understand the undulations in oligotrophic and copiotrophic groups which may otherwise be missed in conventional culturable methods.

3.2 Holobionts

Plant-associated microbes play inimitable roles in nutrient cycling, stress mitigation, disease suppression and pest control along with the improvement of soil properties. In the due course, microbes co-evolved with the associated plants and were mentioned as plant's second genome [6]. The plant system along with its associated microbiome can be collectively termed as holobionts. Although the term holobiont (host and its symbiont) was coined by Lynn Margulis in 1991 and explained as physical association between living organisms belonging to different species, recently it is being commonly used to depict plant–microbe and human-microbe symbiosis [86]. More interestingly, the genome of the host and its microbiota are collectively termed as hologenome. The variations in associated microbial diversity severely influence the evolution of the holobiont [123]. In the perspective of holobiont theory, plant and the associated microbes are considered as single entity and thereby thorough understanding of the functionality of core microbiome of paddy becomes mandatory to maintain crop sustainability.

3.3 Factors Influencing Paddy Microbiome

Primary factor that makes paddy a predominant hub for diversified microbiota is the prevalence of oxic-anoxic interface [26]. Universally, paddy ecosystem is classified as lowland and upland cropping based on the presence or absence of water logging. Diversified bacterial richness has been documented in the former compared to later which substantiates the significant influence of flooding over rice microbiota [19]. Next important factor that impacts paddy microbiome is the availability of different metals and nutrients in cultivated soil. The impact of each individual metal or nutrient over paddy microbial diversity is discussed elsewhere [80, 87]. Furthermore, fertilization imparts mammoth effects over the diversity and functionality of operational taxonomic units (OTU) in paddy ecosystem. In comparison with all soil minerals, nitrogen is hugely dumped into paddy soil due to anthropogenic activity and decoding its effect on plant associated bacteria and archaea population sounds mandatory. Inorganic fertilizer amendment does not significantly alter the core microbiota while addition of N alone hugely modifies their taxonomic diversity [56]. Importantly, paddy-associated microbes including bacteria and fungi significantly accumulate the added N inputs despite low availability [77].

However, integrated nutrient management (INM) including inorganic fertilizers with organic amendments surpasses other fertilization methods by enhancing both soil organic carbon and microbial biomass carbon [37]. After the harvest, paddy straw is incorporated into the field which undergoes multi-stage anoxic degradation into plant available nutrients. In such condition, fertilization with organic amendments rapidly declines the oligotrophs while shooting up the OTU of copiotrophic straw decomposers thereby determining a well-organized bacterial decomposers compared to non-fertilized or synthetic fertilizer amended soil [110]. In yet other outlook, it is claimed that rather than different fertilization, upland-rotation creates a considerable shift in paddy microbiome promoting soil sustainability [40]. Additionally, there is an interesting documentation that soil variables such as pH imparts significant effect on microbial diversity rather than geographic distance [45]. A continental-scale analyses also revealed a positive correlation between chemo-diversity of dissolved soil organic matter and microbial taxonomic profile wherein both play a significant role in soil biogeochemical cycle [59]. However, the crop itself capable of recruiting the associated microbiome, shapes the root-associated bacteria and archaea based on its stage of growth. It is reported that paddy root is colonized by diversified microbial communities in the early stage. While, few taxa outcompetes all other communities and colonize paddy rhizosphere in the later stages based on the selection by the host rather than above listed factors [28]. In a nutshell, paddy microbiota is predominantly decided by soil physicochemical variables, fertilization, geographic distance, age of the plant and upland-lowland rotation systems.

4 Organic Versus Inorganic N Supplements

Application of urea and ammonium sulfate contributes to the N source for paddy fields in most of the countries as paddy prefers NH_4^+ more than NO_3^- [29]. The applied organic manures also get converted into ammonia through mineralization mediated by soil associated microbes. Organic amendments including biochar of bamboo and paddy straw are also recommended to retain N in paddy field by preventing its leaching under water logged condition [27]. While listing the organic sources of N in paddy field, the contribution of the aquatic fern, Azolla and its associative bacteria *Anabaena azollae* becomes inevitable. In general, rice ecosystem emits CH_4 and N_2O while serving as a huge depository for CO_2. Apart from their primary role in fixation of atmospheric N, Azolla cover also decreases the emission of methane from paddy field possibly by increasing the concentration of dissolved oxygen [54]. Paddy-associated microbes play an inevitable role in C and N biogeochemical cycling so as to maintain a positive carbon and nitrogen budget in the soil. Rice ecosystem is also amended with diversified organic supplements which exerts varying effects on paddy microbiota in correlation with global warming. A higher load of heterotrophic bacterial, fungal, actinobacterial and denitrifying bacterial populations were evident in green manure-amended soil followed by soils that received integrated nutrient with organic and inorganic fertilizer [73]. For instance, optimum application of inorganic N fertilizer significantly declined CH_4 emission, while amendment of wheat straw raised up its emission considerably contributing to global warming [64]. Foresaid results were re-validated by documenting increase in the population of methanogenic archaea on rice straw incorporation and rapid reduction in their count on addition of nitrate based fertilizer [4]. In a crisp, CH_4 emission from paddy straw decomposition can be regulated through supplementing optimum dose of N fertilizer, thereby suppressing global warming. In our studies, methanogenic archaea were inhibited under flooded rice soils, owing to competition for H_2 with iron and sulphate reducers and denitrifiers, however, addition of inorganic fertilizers acted as sources of electron donors thus improving the methanogenic flora and methane production [73]. While the methane oxidizers in oxic-anoxic interface, uptakes methane escaped from anaerobic degradation of organic matter [12].

In north-eastern states of India, the paddy residues are burned resulting in waste of essential nutrients apart from causing environmental pollution [62]. In such cases, it is mandatory to create awareness and incline the farmers towards conversion agriculture and make the best N supplement out of paddy straw. Additionally, returning paddy straw to the field creates huge shift in paddy-associated anaerobic microbiome by decreasing denitrifiers and encouraging the population of iron reducers [100]. Paddy straw decomposition is rapidly carried out by hydrogenotrophic and acetoclastic methanogenesis involving diversified bacteria and archaea of various phyla [103]. Reconstruction of paddy associated soil microbial community is also achieved by regular rotation with green manures. Long term rice-rice-green manure rotation has increased the abundance of bacterial taxa such as *Acinetobacter* and *Pseudomonas* which exerts bio-control and other plant beneficial traits [113]. Apart from soil fertility, addition of compost in certain cases has positively enriched the nutrient concentrations in rice [81]. Although positive attributes of organic supplements are noteworthy, heavy application of compost severely reduces soil bacterial diversity despite improving soil fertility and plant nutrition [92]. Recently, the employment of industrial wastes (fly ash, phosphor-gypsum and slag) along with agricultural bi-products (straw, baggase and biochar) are recommended as paddy fertilizers [106]. In this aspect, use of steel slag coupled with biochar has documented significant C sequestration by reducing methane emission in rice field [2, 101]. The farmers of Cauvery delta in Tamil Nadu, India have a practice of applying river sediments to rice paddies every alternate years which may be attributed to the same. Although supplementation of organic and inorganic fertilizers imparts varying effects on core microbiome, most of the studies conclude that combined application of balanced chemical fertilizers with organic amendments would be the best way to shape the microbial community and their functional diversity towards paddy sustainability [26].

5 Impact of Optimum N Fertilization Over Paddy Microflora

Although public agitations on chemical fertilizers are rising up, scientific results emphasis the positive effect of optimum N fertilization on paddy microbial community structure. Low N application coupled with high N use efficiency is achieved by one time addition of zeolite along with inorganic N amendments [105]. Balanced N (ammonium nitrate) fertilization has recorded increased rhizo-carbon deposition by paddy roots thereby recruiting higher diversity of soil microbiome than the plant grown without fertilizer application. Specifically, the population of Gram positive bacteria in rhizosphere and bulk soil significantly increased on stable application of ammonium nitrate based fertilizers [61]. Application of inorganic fertilizers rapidly increased the potential nitrification rate in paddy ecosystem by rising up ammonia oxidizing bacteria and archaea in comparison with N and NK applications [61]. The significance of balanced N and P for regularity of nitrification process is evident from the above said report. Yet another report depicts the synergistic effect of long-term balanced chemical fertilizer and compost application in improving C sequestration in light and heavy fraction of Korean paddy fields [52]. Remarkably, it has been identified that optimum application of ammonium sulphate or lime nitrogen during fallow season, improves paddy straw decomposition by shooting up the population of decomposing microbes. This facilitates stable release of nutrients from paddy straw thereby contributing to high soil fertility prior to next cropping season [90]. As described by Fierer et al. [31] , N fertilization exhibits little effect over bacterial diversity but major effect over their community composition ultimately creating a positive phylogenetic shift in paddy microbiota. However, optimal application of N based chemical fertilizers exerts many positive effects over microbial community structure and their functionality.

6 After-Effects of Imbalanced N Fertilization Over Paddy Microflora

Agricultural systems globally prefer high fertilizer responsive varieties to achieve higher grain yields to feed the growing population. This attitude gradually led to dumping of huge quantity of fertilizers to rapidly increase the yield potential of food crops such as rice. Imbalanced application of N fertilizers, more specifically the higher dose may severely impart negative effects over paddy associated microbial life and pollutes the soil and water ecosystem. China being the largest producer of rice has reported that application of N fertilizers in most of the rice field exceeds the environmentally optimum N rate by 45% and the rate of N application is 10% higher than the rate required for highest yield attributes [48]. Analysis of bacterial community structure in long term fertilized paddy soil, continuously supplied with N fertilizers for past 4 decades concluded that certain beneficial bacterial phyla (Proteobacteria, Acidobacteria and Cyanobacteria) were suppressed, severely affecting soil biogeochemical cycling and rice productivity [56]. Availability of N in soil exerts indirect control over C allocation, paddy rhizodeposits and their uptake by microbiome [33]. Foresaid information warns about the negative effect of excessive N fertilization over C sequestration in paddy soil apart from environmental pollution. Additionally, frequent input of nitrogen fertilizers and their sedimentation in wetland ecosystem predominantly arrests the growth of nitrogen-fixing bacteria and subsequently reduces the nitrogen fixing rate. This process could alter the source of nitrogen for vascular plants in wetland ecosystem that otherwise depends on N fixed by the microbes into their root system [66]. As N input could significantly alter both N and C cycling in paddy ecosystem, cautioned application of optimum dose of fertilizers would help to maintain a healthy biogeochemical cycle along with stable microbiome and sustainable crop yield.

7 Soil Health and Sustainability

N fertilization effect on rice paddies were much focused on their relation with yield and productivity. Their application is not only correlated with their improvement in crop yield, but also has indirect effect on soil physical, chemical and biological properties. Thereby the sustained uses ultimately have pronounced effect on quality and fertility of the soil [116]. The inappropriate usage results in volatilization, leaching, decreased nitrogen use efficiency and so on. Alam and Jia [1] have elucidated the enhanced nitrification activity upon application of urea and ammonium sulphate. Similar results were shown by Verhamme et al. [99] with positive correlation between nitrification activity and ammonium concentration. Atere et al. [3] have observed the interaction effect of N fertilizer and water regime which has effect on rice shoot biomass and photosynthates. They have also hypothesized the stabilized C rhizodeposition upon N treatment in alternate wetting and drying. The assimilate deposition is enhanced in silt and clay fractions, microaggregates and macroaggregates of rhizosphere. Optimizing N fertilization with suitable management practices helps to augment C sequestration. Jiang et al. [46] have studied the positive correlation between GHG emission and applied N fertilizer. Increased carbon footprint was observed due to increased N fertilizer rate. However peak emission of methane were found in medium N application which gradually decreased with higher N. Optimizing N fertilizer input, not only serves as a significant strategy for mitigation of GHG emissions, but also in reducing carbon footprint. Mineralization of N and C strengthen soil fertility and sustainability by various geochemical processes that are triggered by inorganic N [51]. Hence a wide-ranging knowledge on transformation of

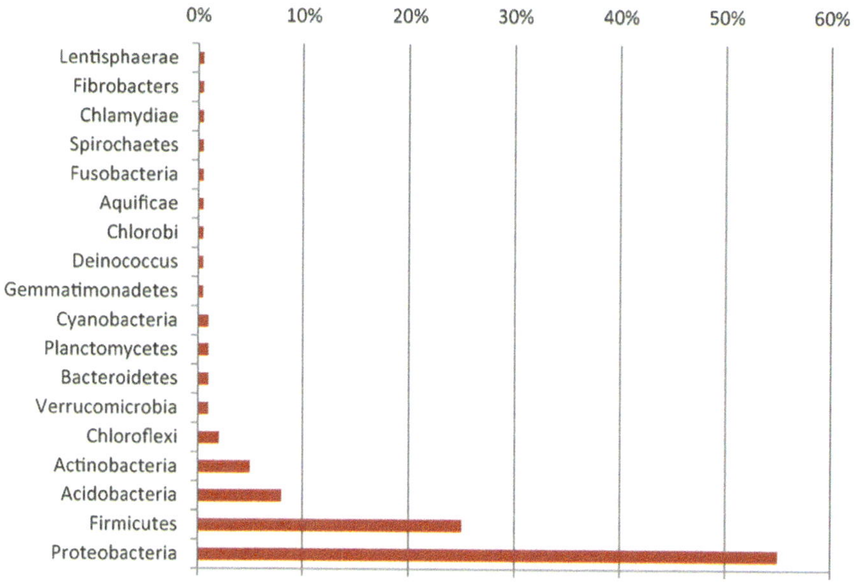

Fig. 1 Phylum level bacterial diversity distribution (in percentage) in tropical flooded rice soil [7]

N in soil in relation to soil organic matter is a mandatory strategy for the improvement of functional microflora of rice and rice productivity (Fig. 1; Tables 1 and 2).

Table 1 N amendments in paddy soil and their consequences on components of N cycling

N fertilization		Soil N transformations				References
		BNF	Ammonification	Nitrification	Denitrification	
Balanced chemical fertilizer		–	+	+	+	[88, 91, 107, 114]
Imbalanced chemical fertilizer	Low N	+	–	–	–	[60, 70, 75, 112]
	High N	–	+	+	+	
Organic amendments		+	+	+	–	[13, 84]
Green manures		+	+	+	+	[16, 54, 98]

Table 2 Impact of N fertilization on microbial genera responsible for N transformation in rice ecosystem

Microbial genera	Role in N transformation	N fertilization type	Effect on bacterial growth	Literature
Bradyrhizobia and *Rhodospirilla*	N fixation	Inorganic N	Enriched growth	[26]
Pseudomonas	Denitrifiaction	Green manure	Enriched growth	[113]
Bacillus	Denitrifiaction	Organic N	Suppressed growth	[100]
Burkholderia and *Enterobacter*	N fixation	Inorganic N	Suppressed growth	[56]
Burkholderia, Bradyrhizobium and *Methylosinus*	N fixation	Inorganic N	Suppressed growth	[41]
Methanogenic archaea	N fixation	Inorganic N	Enriched growth	[41]
Nitrospira	Nitrification	Inorganic N	Suppressed growth	[60]
Nitrosomonas	Nitrification	Inorganic N	Enriched growth	[102]
Nitrososphaera and *Nitrospira*	Nitrification	Combined organic and inorganic N	Enriched growth	[36]
Methanosaeta	Denitrification	Inorganic N	Enriched growth	[96]
Methylobacter, Methylomicrobium, Methylococcus	Denitrification	Inorganic N	Suppressed growth	[1]

References

1. Alam, M., Jia, Z.: Inhibition of methane oxidation by nitrogenous fertilizers in a paddy soil. Front. Microbiol. **3** (2012) https://doi.org/10.3389/fmicb.2012.00246
2. Ali, M.A., Lee, C.H., Kim, S.Y., Kim, P.J.: Effect of industrial by-products containing electron acceptors on mitigating methane emission during rice cultivation. Waste Manage. **29**, 2759–2764 (2009). https://doi.org/10.1016/j.wasman.2009.05.018
3. Atere, C.T., et al.: Rice rhizodeposition and carbon stabilisation in paddy soil are regulated via drying-rewetting cycles and nitrogen fertilisation. Biol Fert Soils **53**, 407–417 (2017). https://doi.org/10.1007/s00374-017-1190-4
4. Bao, Q., Huang, Y., Wang, F., Nie, S., Nicol, G.W., Yao, H., Ding, L.: Effect of nitrogen fertilizer and/or rice straw amendment on methanogenic archaeal communities and methane production from a rice paddy soil. Appl. Microbiol. Biotechnol. **100**, 5989–5998 (2016). https://doi.org/10.1007/s00253-016-7377-z
5. Bei, Z., et al.: Effects of N application strategies on N leaching loss in paddy soil and N use characteristics in different super hybrid rice cultivars. Paddy Water Environ. (2019). https://doi.org/10.1007/s10333-019-00762-x

6. Berendsen, R.L., Pieterse, C.M.J., Bakker, P.A.H.M.: The rhizosphere microbiome and plant health. Trends Plant Sci. **17**, 478–486 (2012). https://doi.org/10.1016/j.tplants.2012.04.001
7. Bhattacharyya, P., et al.: Elucidation of rice rhizosphere metagenome in relation to methane and nitrogen metabolism under elevated carbon dioxide and temperature using whole genome metagenomic approach. Sci. Total Environ. **542**, 886–898 (2016). https://doi.org/10.1016/j.scitotenv.2015.10.154
8. Bodelier, P., Dedysh, S.: Microbiology of wetlands. Front. Microbiol. **4** (2013). https://doi.org/10.3389/fmicb.2013.00079
9. Braker, G., Zhou, J., Wu, L., Devol, A.H., Tiedje, J.M.: Nitrite Reductase Genes *nirK* and *nirS* as functional markers to investigate diversity of denitrifying bacteria in Pacific Northwest Marine sediment communities. Appl. Environ. Microbiol. **66**, 2096–2104 (2000). https://doi.org/10.1128/aem.66.5.2096-2104
10. Buresh, R.J., Casselman, M.E., Patrick, W.H.: Nitrogen fixation in flooded soil systems, a review. In: Brady, N.C. (ed.), Advances in Agronomy, vol 33, pp. 149–192. Academic Press (1980). https://doi.org/10.1016/S0065-2113(08)60166-2
11. Buresh, R.J., Reddy, K.R., Van Kessel, C.: Nitrogen Transformations in Submerged Soils Nitrogen in Agricultural Systems, pp. 401–436 (2008)
12. Cai, Y., Zheng, Y., Bodelier, P.L.E., Conrad, R., Jia, Z.: Conventional methanotrophs are responsible for atmospheric methane oxidation in paddy soils . Nat. Commun. **7**, 11728 (2016). https://doi.org/10.1038/ncomms11728
13. Cao, Y., Sun, H., Zhang, J., Chen, G., Zhu, H., Zhou, S., Xiao, H.: Effects of wheat straw addition on dynamics and fate of nitrogen applied to paddy soils. Soil Till Res. **178**, 92–98 (2018). https://doi.org/10.1016/j.still.2017.12.023
14. Che, S. -g, et al.: Review grain yield and nitrogen use efficiency in rice production regions in China . J. Integr. Agric. **14**, 2456–2466 (2015). https://doi.org/10.1016/S2095-3119(15)61228-X
15. Chen, Z., Luo, X., Hu, R., Wu, M., Wu, J., Wei, W.: Impact of long-term fertilization on the composition of denitrifier communities based on nitrite reductase analyses in a paddy soil. Microbial. Ecol. **60**, 850–861 (2010). https://doi.org/10.1007/s00248-010-9700-z
16. Choudhury, A.T.M.A., Kennedy, I.R.: Prospects and potentials for systems of biological nitrogen fixation in sustainable rice production. Biol. Fert. Soils **39**, 219–227 (2004). https://doi.org/10.1007/s00374-003-0706-2
17. Costa, E., Pérez, J., Kreft, J.U.: Why is metabolic labour divided in nitrification? Trends Microbiol. **14**, 213–219 (2006). https://doi.org/10.1016/j.tim.2006.03.006
18. Cucu, M.A., Said-Pullicino, D., Maurino, V., Bonifacio, E., Romani, M., Celi, L.: Influence of redox conditions and rice straw incorporation on nitrogen availability in fertilized paddy soils. Biol. Fert. Soils **50**, 755–764 (2014). https://doi.org/10.1007/s00374-013-0893-4
19. Cui, H.-L., Duan, G.-L., Zhang, H., Cheng, W., Zhu, Y.-G.: Microbiota in non-flooded and flooded rice culms. FEMS Microbiol. Ecol. **95** (2019). https://doi.org/10.1093/femsec/fiz036
20. Daims, H., et al.: Complete nitrification by Nitrospira bacteria. Nature **528**, 504 (2015). https://doi.org/10.1038/nature16461
21. Daims, H., Lücker, S., Wagner, M.: A New Perspective on Microbes Formerly Known as Nitrite-Oxidizing Bacteria. Trends Microbiol. **24**, 699–712 (2016). https://doi.org/10.1016/j.tim.2016.05.004
22. Delgado-Baquerizo, M., et al.: Microbial diversity drives multifunctionality in terrestrial ecosystems. Nat. Commun. **7**, 10541 (2016). 10.1038/ncomms10541
23. Devêvre, O.C., Horwáth, W.R.: Stabilization of fertilizer nitrogen-15 into humic substances in aerobic vs waterlogged soil following straw incorporation. Soil Sci. Soc. Am. J. **65**, 499–510 (2001). 10.2136/sssaj2001.652499x
24. Dharmappa, P.M., et al.: Introgression of root and water use efficiency traits enhances water productivity: an evidence for physiological breeding in rice (*Oryza sativa* L.) Rice **12**(14) (2019) https://doi.org/10.1186/s12284-019-0268-z
25. Ding, L.J., An, X.L., Li, S., Zhang, G.L., Zhu, Y.G.: Nitrogen loss through anaerobic ammonium oxidation coupled to iron reduction from paddy soils in a chronosequence. Environ. Sci. Technol. **48**, 10641–10647 (2014). https://doi.org/10.1021/es503113s

26. Ding, L.J., Su, J.Q., Sun, G.X., Wu, J.S., Wei, W.X.: Increased microbial functional diversity under long-term organic and integrated fertilization in a paddy soil. Appl. Microbiol. Biotechnol. **102**, 1969–1982 (2018). https://doi.org/10.1007/s00253-017-8704-8
27. Dong, D., Feng, Q., McGrouther, K., Yang, M., Wang, H., Wu, W.: Effects of biochar amendment on rice growth and nitrogen retention in a waterlogged paddy field. J. Soils Sediments **15**, 153–162 (2015). https://doi.org/10.1007/s11368-014-0984-3
28. Edwards, J., et al.: Structure, variation, and assembly of the root-associated microbiomes of rice PNAS 112:E911-E920 (2015). https://doi.org/10.1073/pnas.1414592112
29. Fageria, N.K., Moreira, A., Coelho, A.M.: Yield and yield components of upland rice as influenced by nitrogen sources. J. Plant Nutr. **34**, 361–370 (2011). https://doi.org/10.1080/01904167.2011.536878
30. Fierer, N., Bradford, M.A., Jackson, R.B.: Toward an ecological classification of soil bacteria. Ecology **88**, 1354–1364 (2007). https://doi.org/10.1890/05-1839
31. Fierer, N., Lauber, C.L., Ramirez, K.S., Zaneveld, J., Bradford, M.A., Knight, R.: Comparative metagenomic, phylogenetic and physiological analyses of soil microbial communities across nitrogen gradients ISME J. **6**, 1007–1017 (2012). https://doi.org/10.1038/ismej.2011.159
32. Frenzel, P., Rothfuss, F., Conrad, R.: Oxygen profiles and methane turnover in a flooded rice microcosm. Biol. Fert. Soils **14**, 84–89 (1992). https://doi.org/10.1007/bf00336255
33. Ge, T., et al.: Rice rhizodeposition and its utilization by microbial groups depends on N fertilization. Biol. Fert. Soils **53**, 37–48 (2017). https://doi.org/10.1007/s00374-016-1155-z
34. Geisseler, D., Linquist, B.A., Lazicki, P.A.: Effect of fertilization on soil microorganisms in paddy rice systems—a meta-analysis. Soil Biol. Biochem. **115**, 452–460 (2017). https://doi.org/10.1016/j.soilbio.2017.09.018
35. Geisseler, D., Scow, K.M.: Long-term effects of mineral fertilizers on soil microorganisms—a review. Soil Biol. Biochem. **75**, 54–63 (2014). https://doi.org/10.1016/j.soilbio.2014.03.023
36. Gu, Y., et al.: Long-term fertilization structures bacterial and archaeal communities along soil depth gradient in a paddy soil. Front. Microbiol. **8** (2017). 10.3389/fmicb.2017.01516
37. Hao, X., Liu, S., Wu, J., Hu, R., Tong, C., Su, Y.: Effect of long-term application of inorganic fertilizer and organic amendments on soil organic matter and microbial biomass in three subtropical paddy soils. Nutr. Cycl. Agroecosys. **81**, 17–24 (2008). https://doi.org/10.1007/s10705-007-9145-z
38. Hayashi, K., Nishimura, S., Yagi, K.: Ammonia volatilization from the surface of a Japanese paddy field during rice cultivation. Soil Sci. Plant Nutr. **52**, 545–555 (2006). https://doi.org/10.1111/j.1747-0765.2006.00053.x
39. He, Y., Yang, S., Xu, J., Wang, Y., Peng, S.: Ammonia volatilization losses from paddy fields under controlled irrigation with different drainage treatments. Sci. World J. **2014**, 7 (2014). https://doi.org/10.1155/2014/417605
40. Hou, P.F., Chien, C.H., Chiang-Hsieh, Y.F., Tseng, K.C., Chow, C.N., Huang, H.J., Chang, W.C.: Paddy-upland rotation for sustainable agriculture with regards to diverse soil microbial community. Sci. Rep. **8**, 7966 (2018). https://doi.org/10.1038/s41598-018-26181-2
41. Ikeda, S., et al.: Low nitrogen fertilization adapts rice root microbiome to low nutrient environment by changing biogeochemical functions. Microbe Environ. **29**, 50–59 (2014). https://doi.org/10.1264/jsme2.ME13110
42. Ishii, S., Ikeda, S., Minamisawa, K., Senoo, K., Nitrogen cycling in rice paddy environments: past achievements and future challenges. Microbe Environ. 1110170328–1110170328 (2009a)
43. Ishii, S., Ikeda, S., Minamisawa, K., Senoo, K.: Nitrogen cycling in rice paddy environments: past achievements and future challenges. Microbe Environ. advpub:1110170328–1110170328 (2009b). https://doi.org/10.1264/jsme2.ME11293
44. Isobe, K., Ohte, N.: Ecological perspectives on microbes involved in N-cycling. Microbe Environ. advpub (2014). https://doi.org/10.1264/jsme2.ME13159
45. Jiang, Y., et al.: Crop rotations alter bacterial and fungal diversity in paddy soils across East Asia. Soil Biol. Biochem. **95**, 250–261 (2016). https://doi.org/10.1016/j.soilbio.2016.01.007

46. Jiang, Z., Zhong, Y., Yang, J., Wu, Y., Li, H., Zheng, L.: Effect of nitrogen fertilizer rates on carbon footprint and ecosystem service of carbon sequestration in rice production. Sci. Total Environ. **670**, 210–217 (2019). https://doi.org/10.1016/j.scitotenv.2019.03.188
47. Jiao, S., Xu, Y., Zhang, J., Hao, X., Lu, Y.: Core microbiota in agricultural soils and their potential associations with nutrient cycling. mSystems **4**, e00313-00318 (2019). https://doi.org/10.1128/mSystems.00313-18
48. Ju, X.T., et al.: Reducing environmental risk by improving N management in intensive Chinese agricultural systems. PNAS **106**, 3041–3046 (2009). https://doi.org/10.1073/pnas.0813417106
49. Ju, X.T., Kou, C.L., Zhang, F.S., Christie, P.: Nitrogen balance and groundwater nitrate contamination: comparison among three intensive cropping systems on the North China Plain. Environ. Pollut. **143**, 117–125 (2006). https://doi.org/10.1016/j.envpol.2005.11.005
50. Keeney, D., Sahrawat, K.: Nitrogen transformations in flooded rice soils. Fertil. Res. **9**, 15–38 (1986)
51. Khalil, M.I., Rahman, M.S., Schmidhalter, U., Olfs, H.-W.: Nitrogen fertilizer–induced mineralization of soil organic C and N in six contrasting soils of Bangladesh. J. Plant Nutrit. Soil Sci. **170**, 210–218 (2007). https://doi.org/10.1002/jpln.200520534
52. Kim, K.S., et al.: Effects of long-term fertilization on light and heavy fractions of soil organic matter in single cropping paddy soils in Korea Korean. J. Soil Sci. Fertil. **51**, 616–625 (2018). https://doi.org/10.7745/KJSSF.2018.51.4.616
53. Kim, Y., Liesack, W.: Differential Assemblage of Functional Units in Paddy Soil. Microbiomes Plos One **10**, e0122221 (2015). https://doi.org/10.1371/journal.pone.0122221
54. Kimani, S.M. et al.: Azolla cover significantly decreased CH4 but not N_2O emissions from flooding rice paddy to atmosphere. Soil Sci. Plant Nutr. **64**, 68–76 (2018). 10.1080/00380768.2017.1399775
55. Könneke, M., Bernhard, A.E., de la Torre, J.R., Walker, C.B., Waterbury, J.B., Stahl, D.A.: Isolation of an autotrophic ammonia-oxidizing marine archaeon. Nature **437**, 543–546 (2005). https://doi.org/10.1038/nature03911
56. Kumar, J., De, N., Meena, R., Sharma, P., Pradhan, A., Chari, G.R.: Long term fertilizer management effect on nutrient dynamics in rainfed rice-lentil system in transect 4 of indo-gangetic plain. Int. J. Plant Soil Sci. **26**, 1–10 (2018). https://doi.org/10.9734/ijpss/2018/v26i630057
57. Kuypers, M.M.M., Marchant, H.K., Kartal, B.: The microbial nitrogen-cycling network. Nat. Rev. Microbiol. **16**, 263 (2018). https://doi.org/10.1038/nrmicro.2018.9
58. Ladha, J.K., Reddy, P.M.: Nitrogen fixation in rice systems: state of knowledge and future prospects. Plant Soil **252**, 151–167 (2003). https://doi.org/10.1023/a:1024175307238
59. Li, H.Y., et al.: The chemo diversity of paddy soil dissolved organic matter correlates with microbial community at continental scales. Microbiome **6**, 187 (2018). https://doi.org/10.1186/s40168-018-0561-x
60. Liu, T., Wang, Z., Wang, S., Zhao, Y., Wright, A.L., Jiang, X.: Responses of ammonia-oxidizers and comammox to different long-term fertilization regimes in a subtropical paddy soil. Euro. J. Soil Biol. **93**, 103087 (2019). https://doi.org/10.1016/j.ejsobi.2019.103087
61. Liu, Y., et al.: Initial utilization of rhizodeposits with rice growth in paddy soils: rhizosphere and N fertilization effects. Geoderma **338**, 30–39 (2019). https://doi.org/10.1016/j.geoderma.2018.11.040
62. Lohan, S.K., et al.: Burning issues of paddy residue management in north-west states of India. Renew. Sustain. Energy Rev. **81**, 693–706 (2018). https://doi.org/10.1016/j.rser.2017.08.057
63. Lu, Y., Abraham, W.R., Conrad, R.: Spatial variation of active microbiota in the rice rhizosphere revealed by in situ stable isotope probing of phospholipid fatty acids. Environ. Microbiol. **9**, 474–481 (2007). https://doi.org/10.1111/j.1462-2920.2006.01164.x
64. Ma, J., Li, X.L., Xu, H., Han, Y., Cai, Z.C., Yagi, K.: Effects of nitrogen fertiliser and wheat straw application on CH_4 and N_2O emissions from a paddy rice field. Soil Res. 45:359–367 (2007). https://doi.org/10.1071/SR07039

65. Mo'allim. A.A., Kamal, M.R., Muhammed, H.H., Mohd Soom, M.A., Mohamed Zawawi, M.A.b., Wayayok, A., Che Man, H.B.: Assessment of Nutrient Leaching in Flooded Paddy Rice Field Experiment Using Hydrus-1D Water 10, 785 (2018)
66. Moseman-Valtierra, S.M., Armaiz-Nolla, K., Levin, L.A.: Wetland response to sedimentation and nitrogen loading: diversification and inhibition of nitrogen-fixing microbes. Ecol. Appl. **20**, 1556–1568 (2010). https://doi.org/10.1890/08-1881.1
67. Nannipieri, P., Eldor, P.: The chemical and functional characterization of soil N and its biotic components. Soil Biol. Biochem. **41**, 2357–2369 (2009). https://doi.org/10.1016/j.soilbio.2009.07.013
68. Nicolaisen, M.H., Risgaard-Petersen, N., Revsbech, N.P., Reichardt, W., Ramsing, N.B.: Nitrification–denitrification dynamics and community structure of ammonia oxidizing bacteria in a high yield irrigated Philippine rice field FEMS. Microbiol. Ecol. **49**, 359–369 (2004). https://doi.org/10.1016/j.femsec.2004.04.015
69. Noll, M., Matthies, D., Frenzel, P., Derakshani, M., Liesack, W.: Succession of bacterial community structure and diversity in a paddy soil oxygen gradient. Environ. Microbiol. **7**, 382–395 (2005). https://doi.org/10.1111/j.1462-2920.2005.00700.x
70. Pandey, A., Suter, H., He, J.Z., Hu, H.W., Chen, D.: Nitrogen addition decreases dissimilatory nitrate reduction to ammonium in rice paddies. Appl. Environ. Microbiol. **84**, e00870-e1818 (2018). https://doi.org/10.1128/aem.00870-18
71. Peng, W., Zeng, Y., Shi, Q., Huang, S.: Responses of rice yield and the fate of fertilizer nitrogen to soil organic carbon Plant. Soil Environ. **63**, 416–421 (2017)
72. Ponnamperuma, F.N.: The chemistry of submerged soils. In: Brady, N.C. (ed.), Advances in Agronomy, vol 24, pp. 29–96. Academic Press (1972). https://doi.org/10.1016/S0065-2113(08)60633-1
73. Preethi, B., Poorniammal, R., Balachandar, D., Karthikeyan, S., Chendrayan, K., Bhattacharyya, P., Adhya, T.K.: Long-term organic nutrient managements foster the biological properties and carbon sequestering capability of a wetland rice soil Arch. Agron. Soil Sci. 59 1607–1624 (2013). 10.1080/03650340.2012.755260
74. Prosnansky, M., Sakakibara, Y., Kuroda, M.: High-rate denitrification and SS rejection by biofilm-electrode reactor (BER) combined with microfiltration. Water Res. **36**, 4801–4810 (2002)
75. Ranatunga, T., Hiramatsu, K., Onishi, T., Ishiguro, Y.: Process of denitrification in flooded rice soils Rev. Agric. Sci. 6, 21–33 (2018). 10.7831/ras.6.21
76. Reddy, K., Patrick, W.: Nitrogen fixation in flooded soil. Soil Sci. **128**, 80–85 (1979)
77. Roth, P.J., et al.: Accumulation of nitrogen and microbial residues during 2000 years of rice paddy and non-paddy soil development in the Yangtze River Delta. China Glob. Change Biol. **17**, 3405–3417 (2011). https://doi.org/10.1111/j.1365-2486.2011.02500.x
78. Sahrawat, K.L.: Redox potential and pH as major drivers of fertility in submerged rice soils: a conceptual framework for management. Commun. Soil Sci. Plant Anal. **46**, 1597–1606 (2015). https://doi.org/10.1080/00103624.2015.1043451
79. Said-Pullicino, D., Cucu, M.A., Sodano, M., Birk, J.J., Glaser, B., Celi, L.: Nitrogen immobilization in paddy soils as affected by redox conditions and rice straw incorporation. Geoderma **228–229**, 44–53 (2014). https://doi.org/10.1016/j.geoderma.2013.06.020
80. Samaddar, S., Chatterjee, P., Truu, J., Anandham, R., Kim, S., Sa, T.: Long-term phosphorus limitation changes the bacterial community structure and functioning in paddy soils. Appl. Soil Ecol. **134**, 111–115 (2019). https://doi.org/10.1016/j.apsoil.2018.10.016
81. Sarwar, G., Schmeisky, H., Hussain, N., Muhammad, S., Tahir, M., Saleem, U.: Variations in nutrient concentrations of wheat and paddy as affected by different levels of compost and chemical fertilizer in normal soil. Pak. J. Bot. **41**, 2403–2410 (2009)
82. Savant, N., De Datta, S.: Nitrogen transformations in wetland rice soils. In: Advances in Agronomy, vol 35, pp. 241–302. Elsevier (1982)
83. Shade, A., Handelsman, J.: Beyond the Venn diagram: the hunt for a core microbiome. Environ. Microbiol. **14**, 4–12 (2012). https://doi.org/10.1111/j.1462-2920.2011.02585.x

84. Shu, W., Pablo, G.P., Jun, Y., Danfeng, H.: Abundance and diversity of nitrogen-fixing bacteria in rhizosphere and bulk paddy soil under different duration of organic management. World J. Microbiol. Biotechnol. **28**, 493–503 (2012). https://doi.org/10.1007/s11274-011-0840-1
85. Shuai, W., Jaffé, P.R.: Anaerobic ammonium oxidation coupled to iron reduction in constructed wetland mesocosms. Sci. Total Environ. **648**, 984–992 (2019). https://doi.org/10.1016/j.scitotenv.2018.08.189
86. Singh, Y., Ahmad, J., Musarrat, J., Ehtesham, N.Z., Hasnain, S.E.: Emerging importance of holobionts in evolution and in probiotics. Gut. Pathogens **5**, 12 (2013). https://doi.org/10.1186/1757-4749-5-12
87. Sullivan, T.S., Gadd, G.M.: Metal bioavailability and the soil microbiome. In: Sparks, D.L. (ed.), Advances in Agronomy, vol 155, pp. 79–120. Academic Press (2019). https://doi.org/10.1016/bs.agron.2019.01.004
88. Sun, L., Lu, Y., Yu, F., Kronzucker, H.J., Shi, W.: Biological nitrification inhibition by rice root exudates and its relationship with nitrogen-use efficiency. New Phytol. **212**, 646–656 (2016). https://doi.org/10.1111/nph.14057
89. Szekeres, S., Kiss, I., Kalman, M., Soares, M.I.M.: Microbial population in a hydrogen-dependent denitrification reactor. Water Res. **36**, 4088–4094 (2002)
90. Takakai, F., et al.: Fate of fertilizer-derived N applied to enhance rice straw decomposition in a paddy field during the fallow season under cool temperature conditions. Agriculture **8**, 50 (2018). https://doi.org/10.3390/agriculture8040050
91. Tan, Z., Hurek, T., Reinhold-Hurek, B.: Effect of N-fertilization, plant genotype and environmental conditions on nifH gene pools in roots of rice. Environ. Microbiol. **5**, 1009–1015 (2003). https://doi.org/10.1046/j.1462-2920.2003.00491.x
92. Tian, G., et al.: Improving rice population productivity by reducing nitrogen rate and increasing plant density. PLoS ONE **12**, e0182310–e0182310 (2017). https://doi.org/10.1371/journal.pone.0182310
93. Tian, W., et al.: Responses of microbial activity, abundance, and community in wheat soil after three years of heavy fertilization with manure-based compost and inorganic nitrogen. Agric. Ecosyst. Environ. **213**, 219–227 (2015). https://doi.org/10.1016/j.agee.2015.08.009
94. Toju, H., et al.: Core microbiomes for sustainable agroecosystems. Nat. Plants **4**, 247–257 (2018). https://doi.org/10.1038/s41477-018-0139-4
95. Tyagi, L., Verma, A., Singh, S.N.: Investigation on temporal variation in methane emission from different rice cultivars under the influence of weeds. Environ. Monitor Assess **93**, 91–101 (2004). https://doi.org/10.1023/B:EMAS.0000016793.13260.6e
96. Vaksmaa, A., van Alen, T.A., Ettwig, K.F., Lupotto, E., Valè, G., Jetten, M.S.M., Lüke, C.: Stratification of diversity and activity of methanogenic and methanotrophic microorganisms in a nitrogen-fertilized italian paddy soil. Front. Microbiol. **8** (2017). https://doi.org/10.3389/fmicb.2017.02127
97. van Kessel, M.A.H.J., et al.: Complete nitrification by a single microorganism. Nature **528**, 555 (2015). https://doi.org/10.1038/nature16459
98. Ventura, W., Watanabe, I.: Green manure production of *Azolla microphylla* and *Sesbania rostrata* and their long-term effects on rice yields and soil fertility. Biol Fert Soils **15**, 241–248 (1993). https://doi.org/10.1007/bf00337207
99. Verhamme, D.T., Prosser, J.I., Nicol, G.W.: Ammonia concentration determines differential growth of ammonia-oxidising archaea and bacteria in soil microcosms. ISME J. **5**, 1067 (2011). https://doi.org/10.1038/ismej.2010.191
100. Wang, N., Chang, Z.Z., Xue, X.M., Yu, J.G., Shi, X.X., Ma, L.Q., Li, H.B.: Biochar decreases nitrogen oxide and enhances methane emissions via altering microbial community composition of anaerobic paddy soil. Sci. Total Environ. **581**, 689–696 (2017). https://doi.org/10.1016/j.scitotenv.2016.12.181
101. Wang, W., Lai, D.Y.F., Abid, A.A., Neogi, S., Xu, X., Wang, C.: Effects of steel slag and biochar incorporation on active soil organic carbon pools in a subtropical paddy. Field Agron. **8**, 135 (2018). https://doi.org/10.3390/agronomy8080135

102. Wang, Y., Ke, X., Wu, L., Lu, Y.: Community composition of ammonia-oxidizing bacteria and archaea in rice field soil as affected by nitrogen fertilization. Syst. Appl. Microbiol. **32**, 27–36 (2009). https://doi.org/10.1016/j.syapm.2008.09.007
103. Wegner, C.E., Liesack, W.: Microbial community dynamics during the early stages of plant polymer breakdown in paddy soil. Environ. Microbiol. **18**, 2825–2842 (2016). https://doi.org/10.1111/1462-2920.12815
104. Welte, C.U., et al.: Nitrate- and nitrite-dependent anaerobic oxidation of methane. Environ. Microbiol. Rep. **8**, 941–955 (2016). https://doi.org/10.1111/1758-2229.12487
105. Wu, Q., Chen, T., Chi, D., Xia, G., Sun, Y., Song, Y.: Increasing nitrogen use efficiency with lower nitrogen application frequencies using zeolite in rice paddy fields. Int. Agrophys. **33**, 263–269 (2019). https://doi.org/10.31545/intagr/109545
106. Yan-bing, H., et al.: A three-season field study on the in-situ remediation of Cd-contaminated paddy soil using lime, two industrial by-products, and a low-Cd-accumulation rice cultivar. Ecotoxicol. Environ. Saf. **136**, 135–141 (2017). https://doi.org/10.1016/j.ecoenv.2016.11.005
107. Yang, X.R., et al.: Potential contribution of Anammox to nitrogen loss from paddy soils in Southern China. Appl. Environ. Microbiol. **81**, 938–947 (2015). https://doi.org/10.1128/aem.02664-14
108. Yang, Y., Meng, T., Qian, X., Zhang, J., Cai, Z.: Evidence for nitrification ability controlling nitrogen use efficiency and N losses via denitrification in paddy soils. Biol. Fert. Soils **53**, 349–356 (2017). https://doi.org/10.1007/s00374-017-1185-1
109. Yoshida, M., Ishii, S., Otsuka, S., Senoo, K.: Temporal shifts in diversity and quantity of nirS and nirK in a rice paddy field soil. Soil Biol. Biochem. **41**, 2044–2051 (2009). https://doi.org/10.1016/j.soilbio.2009.07.012
110. Zhan, Y., et al.: Fertilization shapes a well-organized community of bacterial decomposers for accelerated paddy straw degradation. Sci. Rep. **8**, 7981 (2018). https://doi.org/10.1038/s41598-018-26375-8
111. Zhang, J., Lan, T., Müller, C., Cai, Z.: Dissimilatory nitrate reduction to ammonium (DNRA) plays an important role in soil nitrogen conservation in neutral and alkaline but not acidic rice soil. J. Soils Sediments **15**, 523–531 (2015). https://doi.org/10.1007/s11368-014-1037-7
112. Zhang, J., et al.: Agricultural land use affects nitrate production and conservation in humid subtropical soils in China. Soil Biol. Biochem. **62**, 107–114 (2013). https://doi.org/10.1016/j.soilbio.2013.03.006
113. Zhang, X., et al.: Thirty-one years of rice-rice-green manure rotations shape the rhizosphere microbial community and enrich beneficial bacteria. Soil Biol. Biochem. **104**, 208–217 (2017). https://doi.org/10.1016/j.soilbio.2016.10.023
114. Zhang, Y., et al.: Soil N transformation mechanisms can effectively conserve N in soil under saturated conditions compared to unsaturated conditions in subtropical China. Biol. Fert. Soils **54**, 495–507 (2018). https://doi.org/10.1007/s00374-018-1276-7
115. Zhao, Z., et al.: Carbon and nitrogen availability in paddy soil affects rice photosynthate allocation, microbial community composition, and priming: combining continuous 13C labeling with PLFA analysis Plant Soil (2018). https://doi.org/10.1007/s11104-018-3873-5
116. Zhong, W.H., Cai, Z.C.: Long-term effects of inorganic fertilizers on microbial biomass and community functional diversity in a paddy soil derived from quaternary red clay. Appl. Soil Ecol. **36**, 84–91 (2007). https://doi.org/10.1016/j.apsoil.2006.12.001
117. Zhou, J., et al.: Consistent effects of nitrogen fertilization on soil bacterial communities in black soils for two crop seasons in China. Sci. Rep. **7**, 3267 (2017). https://doi.org/10.1038/s41598-017-03539-6
118. Zhou, L., Wang, Y., Long, X.E., Guo, J., Zhu, G.: High abundance and diversity of nitrite-dependent anaerobic methane-oxidizing bacteria in a paddy field profile. FEMS Microbiol. Lett. **360**, 33–41 (2014). https://doi.org/10.1111/1574-6968.12567
119. Zhou, S., Borjigin, S., Riya, S., Terada, A., Hosomi, M.: The relationship between anammox and denitrification in the sediment of an inland river. Sci. Total Environ. **490**, 1029–1036 (2014). https://doi.org/10.1016/j.scitotenv.2014.05.096

120. Zhu, G., Wang, S., Wang, Y., Wang, C., Risgaard-Petersen, N., Jetten, M.S.M., Yin, C.: Anaerobic ammonia oxidation in a fertilized paddy soil. ISME J. **5**, 1905 (2011). https://doi.org/10.1038/ismej.2011.63
121. Zhu, T., Zhang, J., Cai, Z.: The contribution of nitrogen transformation processes to total N2O emissions from soils used for intensive vegetable cultivation. Plant Soil **343**, 313–327 (2011b). https://doi.org/10.1007/s11104-011-0720-3
122. Zhu, Z.L., Chen, D.L.: Nitrogen fertilizer use in China—contributions to food production, impacts on the environment and best management strategies. Nutr. Cycl. Agroecosyst. **63**, 117–127 (2002). https://doi.org/10.1023/a:1021107026067
123. Zilber-Rosenberg, I., Rosenberg, E.: Role of microorganisms in the evolution of animals and plants: the hologenome theory of evolution. FEMS Microbiol. Rev. **32**, 723–735 (2008). https://doi.org/10.1111/j.1574-6976.2008.00123.x

Soil Fauna Activities in Agricultural Greek Landscapes

Evangelia Vavoulidou, Gero Benckiser, and Victor A. Kavvadias

1 Soil Fauna Roles in Soil Ecosystems

Land-use intensification is associated with massive biomass (67%), and biodiversity declines (78%) [1, 2]. In annually grassland samples it was found that the arthropod species decreased in abundance and number of species consistent across trophic levels and independent of the local land-use intensity by 34%. The species specific roles in nutrient cycling and of the soil animal biodiversity, both important dimensions for conservation, are not yet accurate estimated and reasonably understood [3, 4]. The Intergovernmental Science–Policy Platform on Biodiversity and Ecosystem Services (IPBES) consequently aims finding a bridge between above and belowground biodiversity by deciding not focussing exclusively on natural science but also including local, indigenous knowledge systems, social and humanities science, and the policy to implement new insights [5–13].

The soil fauna feeding traits—*detritivorous* or eating dead plant and animal matter and *geophagous* or eating soil—ensure a sustained C, N cycling and soil fertility. Soil habitats, in which protozoans, nematodes, enchytraeids or potworms, soil mites, springtails, insect larvae, Diplopods, Centipedes, woodlice, moles, and earthworms reside, are three-dimensional in structure. The soils inhabiting soil fauna

E. Vavoulidou (✉) · V. A. Kavvadias
Institute of Soil and Water Resources, Department of Soil Science of Athens, Hellenic Agricultural Organization-DEMETER, 1 S. Venizelou Str, 141 23 Lykovrisi, Attiki, Greece
e-mail: evavoul@otenet.gr

V. A. Kavvadias
e-mail: vkavvadias.kal@nagref.gr

G. Benckiser
Institute of Applied Microbiology, Justus-Liebig University, Heinrich Buff-Ring 26-32, Giessen, Germany
e-mail: gero.benckiser@umwelt.uni-giessen.de

G. Benckiser (ed.), *Soil and Recycling Management in the Anthropocene Era*, Environmental Science and Engineering,
https://doi.org/10.1007/978-3-030-51886-8_4

must consume was locally is available, primarily organic matter, depends largely on the unseen microbial majority, which in grassland, savannah soils is responsible for about 5–20% and in temperate and boreal forests for about 80% nitrogen (N) demand supply concerned, and in particular metabolic active earthworms suffer under the intensification of soil cultivation (Fig. 1; Tables 1 and 2; [14, 15]). Earthworms, e.g. *Lumbricus rubellus*, pass organic, clay rich soil, on which microbes are attached, through the gastrointestinal tract and excrete with the faeces microbial survivors, which find in burrow casts good living conditions, increase in abundance, and can feedback on the earthworm health by pre-digesting organic matter and solubilising inorganic soil particles [16]. Microbes, including mycorrhizal fungi are passing the gastrointestinal tract of soil faunal members (Table 1), are thus controlled in multiplication but also spread in the soil environment and can contribute as mycorrhizal fungi to satisfy the phosphorus (P) demand of soil turning around faunal groups by about 75% (Chaps. 2 and 3; [14, 17–29]).

Tillage, climate, and soil faunal activities co-shape the soil texture and soil profiles memorize and illustrate related footprints in quantity and quality (Fig. 1) [29, 32–34]. The soil inhabiting microbial and faunal biodiversity is continuously forced to adapt on the changing availability of soil resources and although ants, moles, and soil components intermixing, for plants nutrients consumable making earthworms (Figs. 2, 3 and 4) comprise only a small fraction within the soil biota biomass (Table 1), they major change a soil profile structure (Fig. 1; [35–38]). Such soil profiles changes give farmers hints how the soil biology could be better integrated in plant productivity increasing cultivation concepts. It is observed across different climates and countries that cultivation intensity directly affects the macro-fauna. In particularly ploughing affects the total earthworm species diversity abundance and biomass, whereas the smaller sized soil fauna members can be favoured [2–4, 39–42]. Also a winter cover crop, as observed at the Horseshoe Bend long-term experimental area in Georgia, United States, can have an influence on the magnitude of earthworms and micro-arthropods [43]. In the winter wheat (*Triticum aestivum* L.) production system abundance and biomass of earthworms and micro-arthropods as collembola (springtails) and mesostigmata mites were under no-tillage higher, whereas conventionally tilled soil favoured the enchytraeid population. Apart from the less sensitive reacting of micro-arthropods on mechanical injury, soil inversion exerted tillage, the trombid forming *Prostigmata* mites seemingly are an exception and the nematode community exhibited cultivation responses concerned conflicting results, which may ascribed to tillage timing and nematode feeding type differences [44–46].

Crotty et al. [49] have studied from 2009 up to 2013 the pivotal role of soil fauna members within different forage crops and land use management after the stony, well-drained loam field, that was either ploughed or direct drilled at sowing, was for 3 years cropped to a stable ryegrass (*Lolium perenne*), perennial red clover (*Trifolium pratense*), white clover (*Trifolium repens*) or chicory (*Cichorium intybus*) system. After the 3 years stable cropping the system was changed an annual arable crop rotation, to a five-year rotation cycle. Tested was in a 4 replicate plots randomised block designed experiment how the soil faunal population alters cultivation dependent in dynamic, diversity and abundance. The first sown crop in 2013 was spring

Widespread soil profiles In Greek landscapes

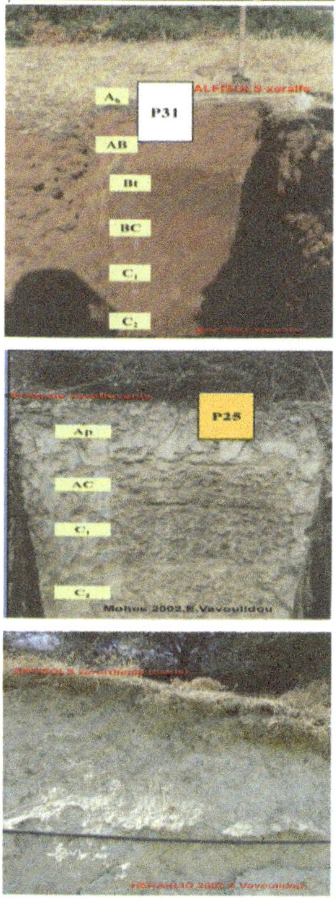

A paddy soil profile

Waterlogged paddy soil,
A-B horizon loam,
C horizon sandy loam ,
Zhe Jiang Province,
Tai-Hu region China
Qi X, Yan-chun L, Yuan-Chang L, Hong-Guan in, China the Tai-Hu region Shanghai, Sci. Tech Publisher, Shanghai

Soil horizon	pH (H_2O)	OM (%)	N (%)	CEC me/100g
A	5.8	4.7	0.24	12.9
P	6.4	4.3	0.23	16.8
WL	7.7	1.1	0.05	12.5
Cg	7.8	-	-	8.6

Anthraquatic paddy soil under long term wheat rotation with a clay-humus in water retaining by wetting, ploughing, harrowing, and 3 important processes shaped soil:

(a) (Fe_3^+/Fe_2^+oxidation-reduction

(b) white bleaching,

(c) formation of Fe-humic coatings (Fe_2^+ to Fe_3^+ oxidation

Soil humus components

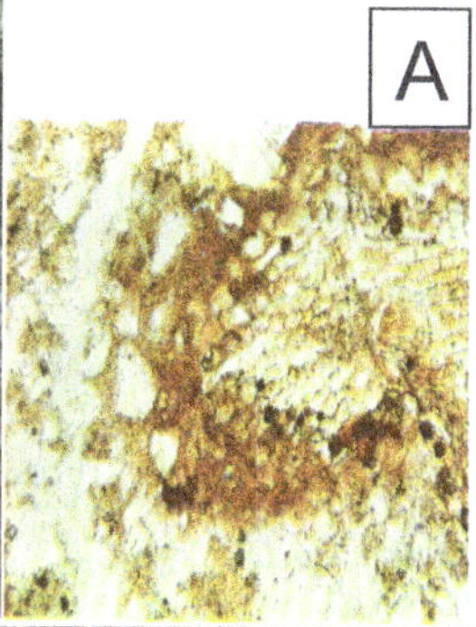

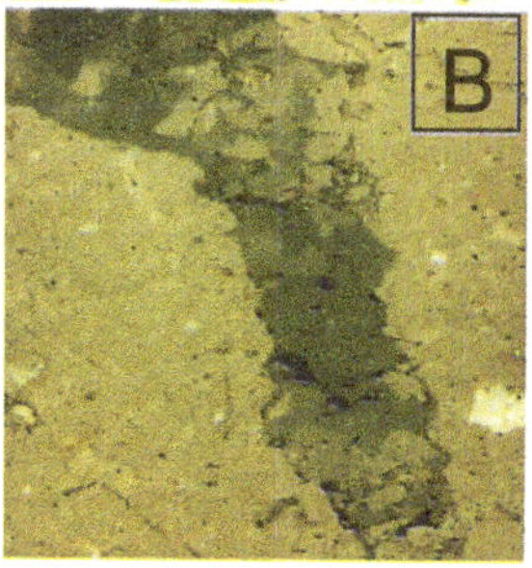

A: Iron oxide encrusted root pseudo-gley, humus horizon (A_h)

B: humus infiltrated sub soil, very likely formed under soil fauna influence

Babel U (1975) Micromorphology of soil organic matter In: Soil components Vol 1, editor Gieseking, chapter7, 378 pages

Iron oxide encrusted root, humus section (bright field, natural size 50 μm). Colourless regions surrounding the iron crusts are quartz grains (30 μm)

Fig. 1 In Greece widespread soil profiles compared with a paddy soil profile, a plant material, intermixed thin section of an A_h and of a loess soil B horizon with infiltrated organic material l, *inter alia* through soil faunal activities, illustrating the process memorizing potential of soil profiles

Table 1 The soil faunal dimensions of biodiversity under a footprint[a]

Taxonomic group	Number of individuals	Biomass (g/m^2)
Bacteria	10^{12}–10^{14}	100–700
Funghi	10^{9}–10^{12}	100–500
Algae	10^{6}–10^{9}	20–150
Protozoa	10^{7}–10^{9}	6–30
Nematodes	10^{4}–10^{6}	5–50
Mites	2×10^{2}–4×10^{3}	0.2–4
Springtails	2×10^{2}–4×10^{3}	0.2–4
Insect larvae	Up to 50	<4.5
Diplopoda	Up to 70	0.5–12.5
Earthworms	Up to 50	30–200

[a]Channarayappa and Biradar [30] and Benckiser [5, 6], Benckiser and Schnell [7]

Table 2 Abundance of earthworms (individuals per m^2) under kinds of crops, tillage systems, and types of fertilization

Tillage system crop	Fertilizer type		
	NP	FYM	Control
Cereals			
Conventional tillage	11.7	17	9
Minimum tillage	17.6	24.3	15.5
No-tillage	27 3	6.3	28
Legumes			
Conventional tillage	20.3	29.2	19
Minimum tillage	25	35	26
No-tillage	46	61	41
Cotton			
Conventional tillage	5.5	9.7	3.7
Minimum tillage	8.3	11	8
No-tillage	13	17	12

NP inorganic nitrogen, phosphorus fertilizer, *FYM* farmyard manure, *Control* without fertilization; Bilalis et al. [31]

wheat (*Triticum aestivum*) and in autumn followed winter barley (*Hordeum vulgare*), harvested in 2014. The overall earthworm abundance and biomass, specifically those of the anecic earthworms (Fig. 4), was higher within the white clover treatment and in the clovers and chicory treatments the nematode functional groups differed with greater numbers of fungal feeders, whilst the greatest abundances of the herbivore nematodes exhibited in the two ryegrass treatments [49]. Although relatively little is known about the links between soil biology, plant type, and cultivation effects, the findings of Croty et al. [49], indicate a relationship between ryegrass and herbivorous

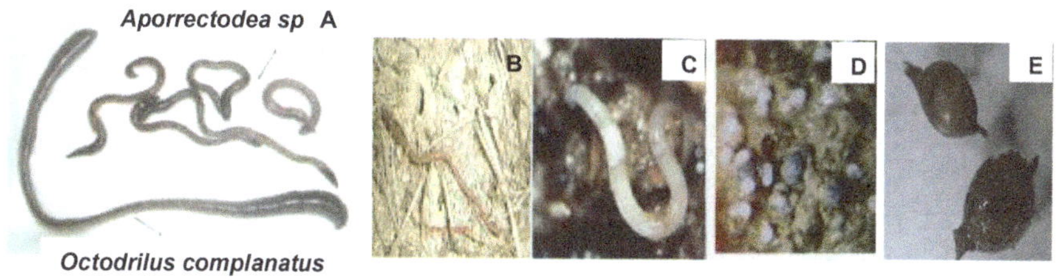

Fig. 2 **a** Typical Greek earthworms in comparison to **b** a *Lumbricidae* earthworm with burrow entry holes and enchytraeids, potworms (genera *Fridericia and Enchytraeus*, **c, d** as representatives of the mesa-fauna phylum Annelida, class Oligochaeta, and **e** cocoons (photos **a, b, c, d, e** Dr. Vavoulidou)

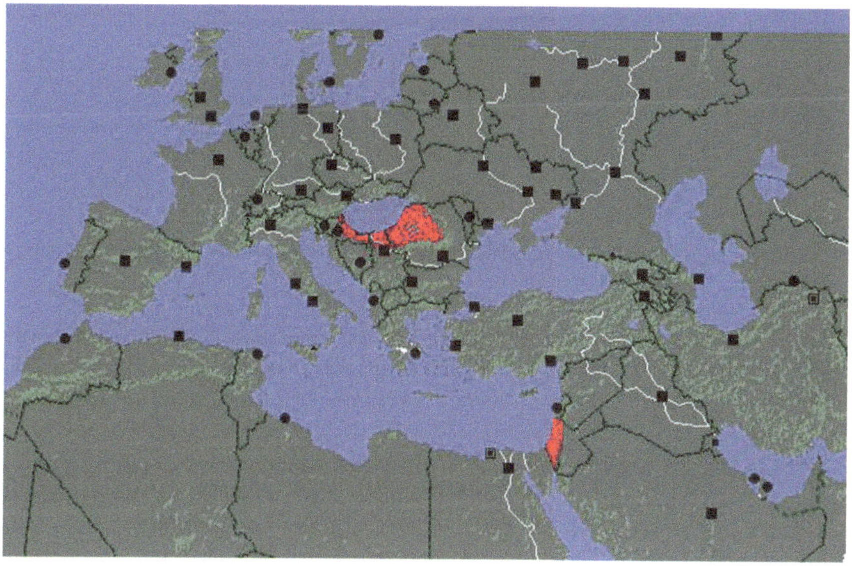

Fig. 3 Earthworms (Oligochaeta: Lumbricidae) in the Palaearctic region of Europa, North-Africa, Asia; exclusively India, the Arabic Peninsula, but inclusively the islands around; Procheş and Syd [47]

invertebrates. Also the observation with the other forages variants provide strong hints that soil food web changes are forage type and cultivation dependent [34, 50–52]. A piecewise proceeding structural equation modelling (SEM), used to test relationships between soil invertebrates response variables and abundance variation, reveals that group of invertebrates may be selectively affected by soil phosphorus availability, soil aggregate stability, soil moisture and in dependency of soil respiration type, the availability of weed or of the produced sorghum sudan grass biomass [53].

The soil fauna members (Table 1) have developed metabolism strategies, selective feeding behaviours, and different morphological shapes allowing to adapt on

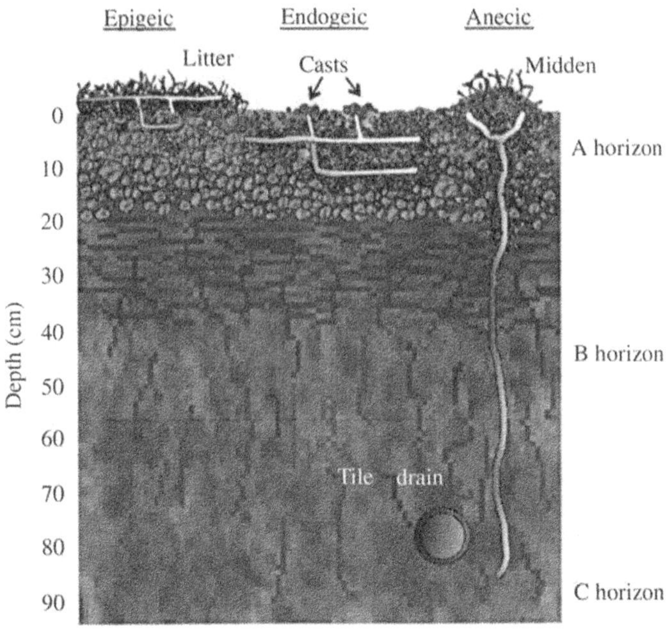

Fig. 4 Burrowing earthworms, picture from Shipitalo [48] Copyright © 2017 by Taylor & Francis

three dimensional soil habitats (Fig. 1). Protozoa, nematodes or collembolan worms consume bacteria, feed on fungal mycelia, or are detritivores [46]. The soil-dwelling mite *Lasioseius fimetorum* Karg (Acari: Podocinidae) feeds on mould mites (*Tyrophagus putrescentiae* [Schrank; Acarina: Acaridae]) and is as the carabid beetle *Nothiophilus biguttatus* a predator. Earthworms prefer various pre-digested litter types by depending on the soil microbial functioning (Chaps. 2 and 3 and by Brown [19], Enkegaard and BrØdsgaard [54–57]). However, a large proportion of microflora-fauna soils co-shaping (Fig. 1) is only poorly understood and in the focus of this chapter.

Conservative terrestrial ecosystem estimates suggest that about 20,000 plant species completely depend on symbiotic interrelationships, on soil fertility. Since the invention of technical N_2 fixation the soil microbial, faunal activities can at least partly be overplayed by an advanced agricultural engineering [58]. A pivotal soil faunal role in the N cycling context is the bacteria feeding of the protozoa and nematodes, because both soil fauna members excrete from the up taken surplus protein not needed ammonia into the surrounding environment [5–7, 59]. Similarly, spread soil turning around ants the waste, majorly in their nests produced by fungus husbandry, around their mounds [60, 61] and burrows with mucus stabilizing earthworms refill the burrows with for microbes nutritionally favourable casts [19, 62]. Each nematode may eat up to 5000 bacteria per minute and digest them, whilst protists may have lower bacteria ingesting rates. The bacteria multiplication controlling nematode community splits in bacteria digesting and in dead organic material preferring groups, what relativizes the soil faunal regulation as also burrowing earthworm exemplify, which not only stabilize the soil bulk density by improving hydraulic conductivity and air permeability, but also destabilize the soil structure (Fig. 1) as soil cultivation and TNF fertilization [10, 13, 18, 63–70]. Soil protists, widespread in the various environments, reside in soil pores, subdivided in highly competitive r- and more environmentally resistant K-selected strategists, in flagellates, amoeba, ciliates, testacea (r = maximal intrinsic rate of natural increase; K refers to carrying capacity; [71]). The around 0.75 mm sized nematodes or roundworms, a representative of the soil micro-fauna is the <20 μm sized*Caenorhabditis elegans*, depend on the soil water content and survive dry periods as cysts. Nemadotes help co-controlling microbial multiplication are thus a keystone in evolution, phylogenetic and functional ecology, in macro-, paleoecology, and conservation biology (Table 1; [12, 44, 49, 66, 72, 73]).

To the mesa-fauna belong the enchytraeids or pot-worms, having a body diameter of 0.05 to 1.5 mm and the smaller sized worms of the phylum Annelida, class Oligochaeta, body length 0.5–20 mm (Fig. 2). Other mesa-fauna members are the until the twentieth century inconspicuous and thus unnoticed proturans, class Entognatha, order Protura, nickname coneheads (soil bears), the wingless, worldwide 12 families and around 830 species comprising soil hexapods, the small, pale, in soil and leaf mould living millipede-like pauropods, the in size larger free-living, humid conditions preferring omnivorous springtails (collembola), the 0.3 mm large Arachnida, *Araneae* with *Nuctenea sclopetaria* as example, and the*Opiliones*, for which earthworms are a significant diet [74–77]. Mesa-fauna members are further the litter inhabiting, 10–180 mm sized Scorpiones (*Euscorpiidae*), of which the females adapt their size on the litter quality, the micro-grazing, micro-predating, detritivore and in nearly all environments found *Isopoda*, the 2–35 cm long detritivore*Diplopoda or* Millipedes with few (around eleven) or over hundred body segments, and the 3–6 cm long *Chilopoda* or Centipedes [14, 78–82].

The order *Hymenoptera* with more than 115,000 described species, the in soils digging ants, termites, moles and earthworms represent the macro-fauna. Earthworms have body sizes 6–30 cm, body widths >2 cm and the family *Lumbricidae* (Fig. 2) comprises 670 known species. The order *Hymenoptera* comprises ants, bees, sawflies, wasps and the closely related order *Isoptera* (termites; [60, 83, 84]). The digging macro-fauna destabilizes during the construction of mounds (nests) or burrows the soil texture and uses nutrients recycling bacteria, archaea, and fungi to pre-decompose litter, which then can be used as food source. Ants, termites and earthworms fertilize their living spaces and the surrounding environment with wastes and excrements and earthworms re-stabilize the soil texture by plugging the burrows with mucus and by condensing the soil during burrow construction [85, 86]. Also the omnivores arachnids, the scavenging and predating *Opiliones* (harvestmen), the moist soil microhabitats inhabiting *Scorpiones* (scorpions), the on soft-bodied fauna feeding *Chilopoda* (centepedes), the on microbially pre-decomposed plant residues feeding *Diploda* (millipedes), woodlice, and isopods are soil macro-fauna members (Table 1). During their digging activities distribute the ecosystem engineers, *Isoptera*, *Hymenoptera*, *Lumbricidae* microflora and the small fauna members in soils by transporting organic residues from the soil surface into the burrows. For farmers are earthworms soil health indicators and not only plants benefit from the service of the about 3500 Annelida species members, the class Oligochaeta (*Annelida: Clitellata*), which *inter alia* help lowering plant fungi intoxication, specifically caused by *Fusarium* species during cereals production, meaning less use of fungicides (Chap. 6; [19, 25, 60, 63, 83, 87]). Earthworms detected in pre-cambrian, odovician sediments are presumably the most ancient macro-fauna soil dwellers, which provide N concerned services of economic value, but suffer under farming intensification, soil compacting devices, and a climate change dependent increased drought and flooding frequency [88–91]. Better models for prediction of soil carbon, nitrogen cycling processes are required, which enable evaluation of soil fauna services, which are the following: (a) enriching by preying and an efficient assimilating of prey the dissolved soil organic matter (DOM) pool with C, N, and P, (b) C/N ratio narrowing of the ingested material to a plant N uptake appropriate 10.4 C/N ratio, and (c) closing the C, N cycling gaps by distributing fertile faeces [90, 92–96].

2 Soil Dwelling Worms and Nitrogen Cycling

The soil ingesting earthworms (*Lumbricidae*; Annelida, Oligochaeta) are 'flagships' among the soil invertebrate groups as worms, isopods, snails, myriapods, mites, springtails (Table 1). They break down large soil and leaf litter particles, help improving as digging ants and moles the soil water conductivity, the aeration of the pore space, and promote in cooperation with the soil microflora, -fauna N cycling and root growth [97–99]. Earthworms increase the litter surface for microbial attacks, intermix organic residues with soil, deposit glycoprotein, polysaccharides, clay enrich casts in the burrows, and when a hectare (ha) farmland receives 100 tons manure per year the soil profiles co-shaping earthworm abundance is high [22, 100–103]. During bioturbation the soil structure physically rearranges and the C/N ratio of dead plant material, transported by earthworms in the burrows narrows from averaged 13 to 10.4 and is then appropriate for plant N uptake. In temperate climate zones earthworms via ingesting plant and soil material, narrow the C/N ratio, may annually deposit 25 to 40 tons faeces per ha, and support the farmland with 28 kg mineralized N/ha [104]. Thereby burrowing rearranges the soil pore system, the water holding capacity and in consequence the mineralizing activity is greater than in the surrounding soil. Because of a benefitting microflora, soil organisms such as springtails, mites and insect larvae earthworms (Fig. 2) became key indicator organisms for detecting soil compaction (Table 2) and chemical, pesticide side-effects [105]. Earthworms are important dietary items for above and belowground living beetles, millipedes, shrews, moles, hedgehogs and omnivorous bird species (blackbirds, thrushes, buzzards, storks and herons). In spite of the profound role the aboveground and soil faunal biodiversity are playing their environmental role is still widely neglected the abundance continually declines, although related studies are initiated [2–4]. In several long-term entomological agronomic research programs, some operate since >170 years, it has been found that due to agricultural intensification the relative earthworm biomass declined by 50–100% (mean 83.3%) in comparison to organic fertilizing [15]. Further a soil moisture reduction by –22.3%, topsoil erosion, a mean loss of SOC/SOM humus by –56.8%, and a 2–9 times earthworm abundance and composition diversity reduction from the optima were recorded. In this context it is observed that recruitment from surrounding fields apparently doesnot balance earthworm declines ([106], [15]). Kopiejewska et al. [25] counted 279–349 individuals (biomass of 103.3–119.5 g) per square meter of a mineral fertilized pasture in comparison to 351–418 earthworm individuals (biomass of 142.4–173.0 g) in an organic fertilized pasture ($P < 0.05$). In both studied pasture treatments the earthworm species *L. terrestris* and *A. rosea* were present, but *Apporectodea caliginosa* dominated the total number by 77.0–93.2% in the mineral fertilized pasture [25]. Across seven of the major biomes on Earth [107] compared in a meta-analysis of >1300 data points the belowground plant, microbial, and faunal biomass. In this study it was found that the microbial biomass carbon represented 0.6–1.1% of soil organic carbon ($r^2 = 0.91$), the belowground plant biomass 1–20% of total carbon ($r^2 = 0.42$), and the belowground soil faunal biomass accounted for <4% of the microbial biomass, for approximately 50% of the total animal biomass across all biomes. In spite of the available >1300 data points our understanding of the global belowground biomass and community structure patterns is limited, although it has been observed that the detritivores and geophagous earthworm "ecosystem engineers", as part of the wide soil microbial/fauna biomass ratio, are majorly active in the upper 15–35 cm soil layer [108, 109]. The databases in the "Acta zoologica Hungaria, 28, 421–454" are currently over 6000 plant litter and mammalian dung consuming, plant material into deeper soil layers transporting, to nutrient turnover contributing, seedling development and plant favouring growth earthworm species, inclusively the megadrile earthworm families suborders (Lumbricina and Moniligastrida), listed and named [110–114]. Identification of juvenile earthworms by morphological comparison is not easy and thus the use of majorly under development DNA based methods is recommended to underestimate not the from soil type to soil type in numbers varying epigeic endogeic and anecic earthworms, particularly in N, P fertilized, limed, and tilled soils (Fig. 4; [115, 116]). Earthworms prefer well- aerated soils in areas with around 800 mm rainfall per year, a good availability of low-humified organic material, a soil pH between 5 and 7, and in such environments the earthworm abundance may range from 30 to 300 individuals (m-3) and the biomass from 110–1100 kg/ha (Table 1; [110]). The maturing periods of growing earthworms is 1–2 years and the lifespan 1.5–4 years while not favourable conditions are related to acidic soils with high clay contents, generally found in waterlogged peats (pH < 4), as well as coarse, freely drained and often too dry soils are not favourable for earthworms [117]. Earthworm species differ in length between 1–1.5 m (weight about 600 g). For example, small, 10–20 mm in size, are forest litter dwelling earthworms, having short lifespans because of the living conditions, whereas for South African earthworm a length of 7 m long (30 kg weight) was reported [118]. In Europe, spread epigeic, endogeic and anecic earthworms (Figs. 3 and 4) are decimated by predating, certain insecticides, and soil tillage, survive between 0–35 °C, ingest soil minerals and sand grains for aiding matter grinding in their gizzard, and stabilize the burrow walls with mucus (Table 2; [119, 120]). Earthworm mucus is a valuable nutrient source for microbes as well as the earthworm faeces, deposited on the soil surface and in the burrows. *Lumbricus terrestris* (Fig. 2) forms deep vertical burrows of up to 3 m with several branched entrances at the surface and near the soil surface living earthworms build horizontal burrows with occasional shafts into the soil to vertical burrows (Fig. 4). The deposited earthworm casts in burrows are rich in nutrients and contain a large number of bacteria, archaea, and viable propagules of mycorrhizal fungi, which after regaining activity produce more plant available nutrients than the surrounding soil [121]. Over a year, earthworms, which can ingest between 8.9 and 182 tons soil and organic material/ha, withstand dry or cold periods in the mucus stabilized burrows, which they leave actively or forced by heavy rains, waterlogging [104]. In the Mediterranean region around Greece less impairing living conditions for earthworms are found during the colder period from October to April; (Table 3). Under better living conditions undecomposed surface residues are integrated into the mineral soil, leading to reduced nutrient losses, development of soil structure, improvement of nutrients availability and soil fertility, and crop yield increases [3].

The European Common Agricultural Policy (CAP) has taking biodiversity into its portfolio and the UN Sustainable Development Goals highlight that soil biota abundance is of great importance for the provision of associated ecosystem services

Table 3 Mean Greek climatic conditions, with a moderate temperature and higher precipitation rates from October on towards April

	Site	Mean precipitation (Pmm)	Mean temperature (Tco)	% Precipitation	Mean temperature Oct–April	Mean temperature
		Annual		Oct–April		May–Sept
Larissa (Tirnavos)	Larissa	466	16.2	72.0	10.2	24.6
Lamia	Lamia	574	16.5	79.1	11.1	24.1
Viotia	Voiotia	489	17.0	82.2	11.7	24.4
Attiki	Attiki	416	17.6	85.5	12.4	24.9
W. Greece-Arta	W. Greece	1035	16.33	86.6	9.6	22.9
N. Pelep., Nemea, Egeio	Nemea	24	17.0	86.8	11.4	23.4
Achaia	Achaia	660	18.3	88.0	15.9	25.0
Ilia	Ilia	818	18.2	87.4	13.4	24.4
Messina	Messinia	1067	15.6	81.0	10.7	23.7
Arkadia	Argadia	896	14.4	83.9	9.7	21.8
Crete	Chania	665	18.7	91.1	14.4	24.4
Thera (volcanic)	Santorini	370	17.5	92.4	13.9	22.6

(ES). Nevertheless, the process regulating belowground biota, among which the crop health protecting earthworms are fundamental drivers of self-regulation, finds little notice and suffers under intensifying agro-environmental measures and an obviously changing climate [98, 122, 123]. Earthworms, shielding the productivity of winter wheat from the toxic plant fungi *Fusarium*, are crucial in soil live balancing, soil water retention potential, soil healthy soils, plant biomass growth, food production security, soil structure stabilisation, and in helping to minimize erosion, desertification, landslides, and climate change mitigation. In the drilosphere, the soil volume under earthworm influence, earthworms may cause drastic shifts in soil density, structure, and microbial and fauna communities diversity, (a) by comminution, burrowing and casting; (b) by ingesting dwelled organic and inorganic soil material; (c) by dispersing the ingested soil life that must adapt on earthworm gut passage [124]. Pauli et al. [125] consulted farmers and asked them what they know concerning soil microflora and fauna communities functioning and experienced an in depth going farmers' knowledge about the invisible, hidden soil microflora and fauna diversity in Earth's soil habitats, but also a large unawareness of the soil fauna potential. Some farmers held detailed observations of local soil biological activity and use biota counts and taxonomies as soil fertility indicators and for getting better insights in soil properties dependent earthworm abundance. For achieving a greater interest in organic cropping systems the Greek Ministry of Agriculture, financially supported by the European Union (EU), could encourage farmers to participate in a 5 years organic farming promoting program. In December 1993 surface soil sampling (0–30 cm) took place in different Greek agriculturally sites (cereal fields, olive and citrus groves, vineyards, various fruit tree plantations, fallow land) in order to study the effect of nitrogenous fertilization and cultivation system on the earthworm abundance [126]. In 316 collected soil samples earthworm abundance, clay, silt, and sand percentage (soil texture), soil pH, organic matter, total N and other nutrients contents were determined as well as NH_4^+ and NO_3^-–N in approximately 50 soil samples [126, 127].

3 The Greek, Cultivation Dependent Abundance of the Earthworm Population

Surface soil samples (0–30 cm) were taken during the months (October to April) across various agricultural areas of Greece (Fig. 5) by considering crop type and cultivation system, and the earthworm abundance in relation to soil physical and chemical properties was estimated, the species and subspecies identified (Fig. 5; Tables 3 and 4; [126, 128]). Soil samples were collected across various agricultural areas of Greece by considering crop type and cultivation system. From the 67 found individual *Clitellata*: *Megadrili*species and subspecies (Table 4) 59 taxa belonged to the *Lumbricidae*, 3 to *Megascolecidae*, 2 to *Acanthodrilidae* and *Ocnerodrilidae*, and 1 taxon to the *Criodrilidae* family. The earthworm species *Eisenia spelaea var.*

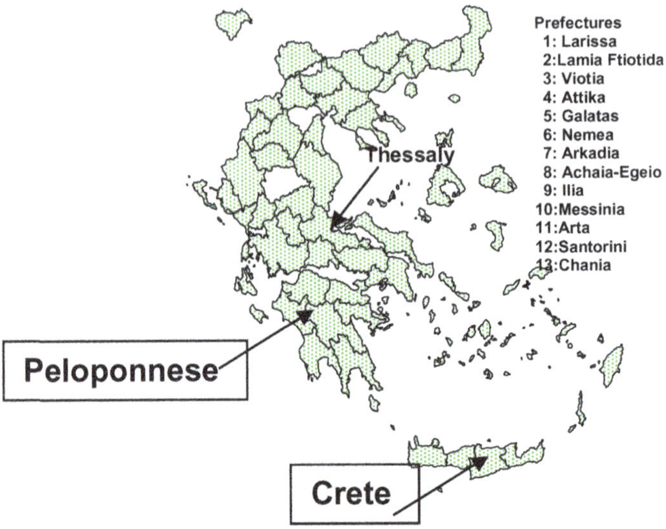

Fig. 5 Soil sampling in more than 150 sites in the landscapes across Greece in which the earthworm abundance was determined between 1998 and 2007 within the on organic farming focussing FAO-program

athenica [129], synonym to *Aporrectodea rosea* ([130], *Allolobophora kosowensis* Karaman 1968, *Amynthas gracilis* [131], and *Eukerria saltensis* (Beddard 1895) (Table 4) were recorded for the first time. In addition 9 from the *Lumbricidae* family of the Greek earthworm fauna were characterized as strict and 10 as Balkanic endemic species [132]. By correlating the earthworm abundance with the concomitantly determined soil properties (Table 5) exhibited a significant positive correlation (F value of 2,571, $P < 0.05$) between earthworm abundance and organic matter availability, but the correlation between total N content and earthworm abundance was not significant. In addition, the correlation between NH_4^+-N and NO_3^-–N and earthworm abundance showed that the earthworm abundance seems to depend on the NH_4^+-availability (p 0.05%). The polysaccharide content, the total and mineral nitrogen level of casts with ammonia as major part, much greater than that of soil, is rapidly converted to nitrate [133] and from a soil sample campaign in organic and conventional treated olive groves in central Greece was reported that the earthworm community was significantly higher in the organic than in the conventional olive groves ($p < 0.05$; [128]). However, the data of Table 5 reveal a higher earthworm abundance in the conventional treatments. According to the olive grove study by 124 majorly the species *Octodrilus complanatus*, *O. croaticus*, *Dendrobaena byblica*, *D. veneta*, *Aporrectodea caliginosa,* and *Microscolex phosphoreus* could be identified while the population of *Lumbricus terrestris* was more affected than *Allolobophora longa*, *A. caliginosa* or *A. chorotica* population and *D. veneta* and *M. phosphoreus* were absent in the conventional treatment ([132, 134, 135]). The largest earthworm populations in the long-term Broadbalk, Barnfield, and Park Grass experiments, continuous grown

Table 4 Chorological groups of the earthworms in Greece (for more details see [132])

Species	Chorological group
Dendrobaena epirotica (Zicsi and Michalis 1993)	Endemism
Dendrobaena mahnerti (Zicsi 1974)	Endemism
Dendrobaena pindonensis (Zicsi and Michalis 1993)	Endemism
Dendrobaena retrosella (Szederjesi and Csuzdi 2012)	Endemism
Eisenia ariadne (Michaelsen 1928)	Endemism
Eisenia oreophila (Szederjesi and Csuzdi 2012)	Endemism?
Allolobophora eurytanica (Tzelepe 1943)?	Endemism
Dendrobaena byblica olympiaca (Michaelsen 1902)	Southern Balkanic endemism
Dendrobaena michalisi (Karaman 1972)	Southern Balkanic endemism
Eisenia ebneri (Michaelsen 1914)	Southern Balkanic endemism
Allolobophora dofleini (Ude 1922)	Balkanic endemism
Allolobophora kosowensis (Karaman 1968)	Balkanic endemism
Dendrobaena balcanica (Černosvitov 1937)	Balkanic endemism
Dendrobaena bokakotorensis (Šapkarev, 1975)	Balkanic endemism
Dendrobaena hrabei (Černosvitov 1934)	Balkanic endemism
Dendrobaena olympica [129]	Balkanic endemism
Dendrobaena rhodopensis (Černosvitov 1937)	Balkanic endemism
Helodrilus balcanicus (Černosvitov 1931)	Balkanic endemism
Helodrilus vagneri (Mršić 1991)	Balkanic endemism
Spermophorodrilus antiquus [129]	Balkanic endemism
Perelia phoebea (Cognetti 1913)	Aegean endemism
Dendrobaena byblica byblica (Rosa 1893)	Circum-Mediterranean
Eiseniella neapolitana (Örley 1885)	Circum-Mediterranean
Murchieona minuscula (Rosa 1905)	Circum-Mediterranean
Octodrilus complanatus (Dugès 1828)	Circum-Mediterranean
Criodrilus lacuum (Hoffmeister 1845)	Circum-Mediterranean
Aporrectodea jassyensis (Michaelsen 1891)	Trans-Aegean
Octodrilus transpadanus (Rosa 1884)	Trans-Aegean
Proctodrilus tuberculatus (Černosvitov 1935)	Trans-Aegean
Dendrobaena pentheri (Rosa 1905)	Caucasian-East-Mediterranean
Eisenia colchidica (Perel 1967)	Caucasian-East-Mediterranean
Helodrilus patriarchalis (Rosa 1893)	Caucasian-East-Mediterranean
Eisenia kattoulasi (Zicsi and Michalis 1981)	Balkanic-Anatolian
Fitzingeria loebli (Zicsi 1985)	Balkanic-Anatolian

(continued)

Table 4 (continued)

Species	Chorological group
Dendrobaena alpina alpina (Rosa 1884)	Balkanic-Alpine
Dendrobaena attemsi (Michaelsen 1902)	Balkanic-Alpine
Proctodrilus antipai (Michaelsen 1891)	Central-European
Proctodrilus opisthoductus (Zicsi 1985)	Central-European
Aporrectodea georgii (Michaelsen 1890)	Atlanto-Mediterranean

to wheat since 1843, to root crops since 1843, and to grass since 1836, respectively, were found in the plots having received both inorganic and organic N fertilizers and a strong positive correlation ($r = 0.9825$) exhibited between applied inorganic N amounts and the earthworm populations [136]. Such a strong correlation not exhibited in Table 4, assumingly because the organic matter and clay rich soil particles ingesting earthworms in the fine textured, often calcareous Entisols and Inceptisols and in the mainly acid Alfisols of Greece, with a most common pH between 5 and 9, could have suffered under bound and taken up trace elements. The endogeic earthworm *Metaphire posthuma*, a sensitive test model for ecotoxicological studies, may weigh 0.5 g on average and produces casts of approximately five times the own weight per day [137, 138]. If such a worm would be active from October to Apri in Greece (Table 3) it could produce during this period around 450 g cast, in which the polysaccharide content is much greater than that of soil [133]. The top soil (30 cm) Cu content in different Greek regions ranges between 16 mg kg^{-1} (Limnos and Santorini) and 89 mg kg^{-1} (Nemea), with an overall mean of 38 ± 26 mg kg^{-1} mg per kg [139]. Exceeds on dry weight basis the calculated Cu concentration within earthworm bodies 100–120 mg kg^{-1} the worm may die [140], meaning at the overall mean Cu content of 38 ± 26 mg per kg an earthworm must uptake around 3 kg soil with the Greek soil Cu content an earthworm as *M posthuma* is impaired, but the pH may play a role, because earthworms have limited abilities to adapt to soil acidity except some acidic soils tolerating species [111, 141–144]. Soil Cu content concerned is to consider that earthworms can escape even moderately toxic situations. Ideal is a neutral pH as found in the calcareous Greek Entisols, Inceptisols. Liming of the partly in acid Alfisols as pH correcting measure would not much improve the earthworm species survival, but nematodes and enchytraeids could markedly decrease in numbers. Liming generally may mean dramatic shifts in the composition and abundance of soil fauna communities, in particular in acid soils [11]. In the island Santorini, an important wine-producing area of Greece, Cu values in the loam sandy volcanic soils were close to the overall Greek average 38 mg Cu kg^{-1} except in the orchard/nut plantation sites, in which of some Cu was as high as 70 mg kg^{-1}. In the loam sandy volcanic soil of Santorini *Lumbricidae* earthworms were widely absent and in conventionally or organically with vines and tomatoes cultivated locations enchytraeids observed in a low abundance (Fig. 6; [139, 145]).

Table 5 Earthworm counts in soil samples taken across Greece (for landscape examples see Fig. 3)

Biotopes	Earthworm Individuals m^{-2}	Organic matter	Nitrogen total	C/N	pH	Clay (%)
Cultivated arable land[a]						
Conventional treated	41 ± 15	1.9 ± 0.1	0.14 ± 0.01	8.1 ± 0.4	7.64 ± 0.1	49 ± 2
Extensive fallow	48 ± 34	2.4 ± 0.4	0.18 ± 0.02	7.9 ± 1.1	7.55 ± 0.25	48 ± 4
Vineyards[b]						
Conventional treated	16 ± 9	1.9 ± 0.2	0.14 ± 0.01	8.0 ± 0.5	7.2 ± 0.12	45 ± 2
Organic farmed	52 ± 23	1.5 ± 0.3	0.1 ± 0.02	8.7 ± 0.5	6.7 ± 0.19	37 ± 3
Extensive fallow	60 ± 17	2.2 ± 0.2	0.13 ± 0.01	10.0 ± 0.5	7.5 ± 0.13	44 ± 2
Abandoned land	106 ± 28	2.7 ± 0.4	0.15 ± 0.02	10.3 ± 0.5	7.0 ± 0.25	33 ± 4
Olives[c]						
Conventional treated	63 ± 29	2.8 ± 0.4	0.19 ± 0.03	8.1 ± 1.3	6.8 ± 0.29	35 ± 5
Organic farmed	36 ± 20	2.6 ± 0.3	0.17 ± 0.02	11.2 ± 0.9	7.0 ± 0.20	36 ± 3
Extensive fallow	82 ± 26	2.5 ± 0.3	0.17 ± 0.02	9.1 ± 0.9	7.4 ± 0.21	44 ± 3
Abandoned land	43 ± 42	5.0 ± 0.5	0.28 ± 0.03	9.5 ± 1.5	7.0 ± 0.33	33 ± 5
Fruit trees[d]						
Conventional treated	34 ± 17	2.2 ± 0.2	0.18 ± 0.01	8.3 ± 0.6	6.7 ± 0.13	37 ± 2
Organic farmed	75 ± 29	2.8 ± 0.4	0.19 ± 0.02	8.6 ± 1.1	7.0 ± 0.25	33 ± 4
Extensive fallow	135 ± 24	3.4 ± 0.3	0.21 ± 0.02	10.0 ± 0.9	7.4 ± 0.19	44 ± 3
Abandoned land	31 ± 48	4.0 ± 0.5	0.20 ± 0.03	11.5 ± 1.6	6.0 ± 0.37	33 ± 6

[a]Cereals (n = 37) + cotton (n = 12) + vegetable (n = 17) + legumes (n = 3) + pastures (n = 4)
[b]Convential (n = 25) + organic (n = 13) + extensive fallow (n = 37) + abandoned land (n = 15)
[c]Convential (n = 4) + organic (n = 15) + extensive fallow (n = 13) + abandoned land (n = 6)
[d]Citrus n = 33: conventional n = 9, organic n = 15, fallow n = 4; Chest-, Wallnut n = 8
[e]Pistacia n = 5; Magnolia n = 2; Oak forest and Macchia n = 6

Fig. 6 Soil cultivation examples of selected Greek in conventional, organic cultivated cereals, cotton, vegetables, and fallow agricultural sites (Central Greece), olives, citrus plantations (West Greece), some fruit tree locations, vineyards (Peloponnese, Crete). Soil properties and earthworm details see Tables 4 and 5

4 Soil Fauna Future Food Production Perspectives

Ecologists around the world profusely explore soils and in recent years evaluation of the belowground organism evolution is in the focus [2–4]. Food web abundances and patterns changes, biodiversity losses across different ecosystems are increasingly observed, soil profiles (Fig. 1) memorize the societal polluting, the impacts of agricultural cultivation systems, and the waste producers and biodiversity losses causer are identified (Chap. 1). In spite of the availability of DNA sequencing methods, which continually are improved, our state of knowledge of the local soil species diversity is still modest and the best available technology in laboratory and field is still the interpreting of soil fertility with earthworms as model [5–7, 35, 105].

Figure 6 provides as example a plants, animals, soil biodiversity, and economy integrating landscape model that is based on an evaluating of their benefits beyond soil quality/health. Biodiversity protecting directives and international commitments, e.g., Habitats Directive, Natura 2000, or Convention on Biological Diversity, CITES), refer to legacy effects across three organism scales (microbes, plants, animals) by not explicitly mentioning the soil fauna biodiversity [18, 146, 147]. Thus, we need not only a further refinement of our experimental approaches, more taxonomical efforts and ecological work on soil biota level, but also "activity proxies" and "response traits" in addition to abundance and biomass estimates. When it is true that soils sustain a high diversity, how could the little attention they receive compared to plants or birds be justified?

Until now we are unable to clearly relate ecosystem functions to ecosystem diversity. Relationships, found in several studies, might not be causal correlations and hence, much of the diversity present might just be "redundant." For preserving plants seed banks have been built in selected parts of the world to keep unique and valuable plant species. For saving microbial genomes large culture collections of fungi (including yeasts), bacteria and plasmids exist for commercial value and for keeping the possibility of "resurrecting" nearly extinct aboveground vertebrate species, cells, tissues and embryos from frozen stored samples [148–150]. In soils (Fig. 1) decaying is organized with involvement of a large belowground biodiversity and for describing their functionality bio-indicators earthworm abundance, nitrogen, soil condition relationship is estimated and discussed [113]. A re-defined humus index, further developed from that of Ponge et al. [151], could provide a soil biodiversity, humus forms integrating framework for soils, plants, animals, and economy integrating landscape models [18]. For providing a re-defined humus index based integrative soil food web framework *inter alia* the question whether burrowing, soil ingesting earthworms are selective feeders or ingest they just accidentally what they eat should clearly be answered. From protozoa, nematode, collembolan, and the epigeic single fungal species preferring *L. rubellus* (Fig. 4) we know that soil inhabitants feed selectively and related research may open new options in elucidating the level of species-specific selection by grazing and assessing the overall impacts of gut passage shaping [11, 152, 153]. The earthworm gut passage selects between survivors and non-survivors and thus have different implications for soil biodiversity attacks on recalcitrant substrates, crop performance in dependency of fungi infestations, and finally on model approaches (Fig. 7), which for constructive mathematical formulations, e.g. should know when seasonal litter fall and nutrients pulses activate the biology in hot spots or when to what extent horizontal and vertical transfers of leaf residues on the soil surface and deeper soil layers take place. The current literature descriptions of the processes at the root-soil interface dominate, but a subordinated attention find soil fauna effects, termed "zoological weathering" (elements mobilizing), "zoological bonding and retarding" (organic matter and biology immobilizing), "zoological bioturbation" (soil structure shaping), modulating the breaking down of organic inputs, the integrating of abiotic (climate, soil texture) and biotic factors (litter quality and biological accessibility), and influence decomposition across spatial and temporal scales.

The complex nature of soil communities, living in a heterogeneous mosaic of microsites with different conditions and resource availabilities where abundance and

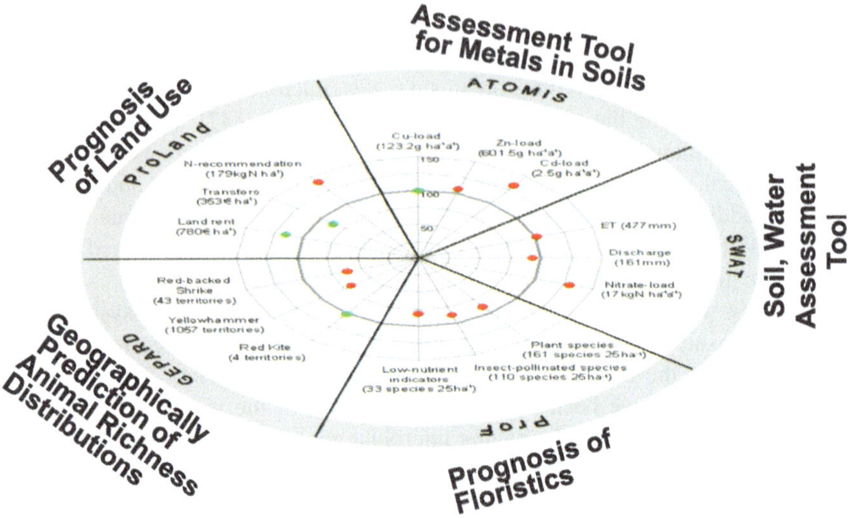

Fig. 7 Ecological–economic supply chain pyramid, or landscape multifunctionality scenarios (MS) integrating model, developed for a sustaining precison-, organic farming in the Wetterau region, Hesse, Germany (red dots: Negative, green dots Positive; Estimated total MS indicator values in brackets. ET, evapotranspiration (Waldhardt et al. Ecol Soc 15, 30. URL: https://www.ecologyandsociety.org/vol15/iss3/art30)

composition of species assemblages determine the temporal dynamics this chapter intends to illustrate. Soils and their inhabitants do not conform to closed loops (Chap. 2, Fig. 2) and it is not easy to develop mechanisms allowing shelf-organization [64, 154]. A more refined view of all the potential contributions of the soil biodiversity organisms to soil processes is needed that International agendas **as** GSBI (Global Soil Biodiversuty Initiative, [155]), IPBES (Intergovernmental Science-Policy Platform on Biodiversity and Ecosystem Services, [156]), or the UN Sustainable Development Goals, can better integrate their efforts in a sustainable management and climate change mitigation policies.

References

1. Coyle, D.R., Nagendrab, U.J., Taylor, M.K., Campbella, J.H., Cunard, C.E., Joslin, A.H., Mundepi, A., Phillipse, C.A., Callaham, Jr, M.A.: Soil fauna responses to natural disturbances, invasive species, and global climate change: current state of the science and a call to action. Soil. Biol. Biochem. **110**, 116–133 (2017). https://doi.org/10.1016/j.soilbio.2017.03.008
2. Seibold, S., Gossner, M.M., Simons, N.K., Blüthgen, N., Müller, J., Ambarlı, D., Ammer, C., Bauhus, J., Fischer, M., Habel, J.C., Linsenmair, K.E., Nauss, T., Penone, C., Prati, D., Schall, P., Schulze, E.D., Vogt, J., Wöllauer, S., Weisser, W.W.: Arthropod decline in grasslands and forests is associated with landscape-level drivers. Nature **574**, 671–674 (2019). https://doi.org/10.1038/s41586-019-1684-3

3. Phillips, H.R.P., Guerra, C.A., Bartz, M.L.C., Briones, M.I., Brown, G., Crowther, T.W., Ferlian, O., Gongalsky, K.B., van den Hoogen, J., Krebs, J., Orgiazzi, A., Routh, D., Schwarz, B., Bach, E.M., Bennett, J., Brose, U., Decaëns, T., König-Ries, B., Loreau, M., Mathieu, J., Mulder, C., van der Putten, W.H., Ramirez, K.S., Rillig, M.C., Russell, D., Rutgers, M., Thakur, M.P., de Vries, F.T., Wall, D.H., Wardle, D.A., Arai, M., Ayuke, F.O., Baker, G.H., Beauséjour, R., Bedano, J.C., Birkhofer, K., Blanchart, E., Blossey, B., Bolger, T., Bradley, R.L., Callaham, M.A., Capowiez, Y., Caulfield, M.E., Choi, A., Crotty, F.V., Dávalos, A., Cosin, D.J.D., Dominguez, A., Duhour, A.E., van Eekeren, N., Emmerling, C., Falco, L.B., Fernández, R., Fonte, S.J., Fragoso, C., Franco, A.L.C., Fugère, M., Fusilero, A.T., Gholami, S., Gundale, M.J., López, M.G., Hackenberger, D.K., Hernández, L.M., Hishi, T., Holdsworth, A.R., Holmstrup, M., Hopfensperger, K.N., Lwanga, E.H., Huhta, V., Hurisso, T.T., Iannone, B.V., Iordache, M., Joschko, M., Kaneko, N., Kanianska, R., Keith, A.M., Kelly, C.A., Kernecker, M.L., Klaminder, J., Koné, A.W., Kooch, Y., Kukkonen, S.T., Lalthanzara, H., Lammel, D.R., Lebedev, I.M., Li, Y., Lidon, J.B.J., Lincoln, N.K., Loss, S.R., Marichal, R., Matula, R., Moos, J.H., Moreno, G., Morón-Ríos, A., Muys, B., Neirynck, J., Norgrove, L., Novo, M., Nuutinen, V., Nuzzo, V., Rahman, M., Pansu, J., Paudel, S., Pérès, G., Pérez-Camacho, L., Piñeiro, R., Ponge, J.F., Rashid, M.I., Rebollo, S., Rodeiro-Iglesias, J., Rodríguez, M.Á., Roth, A.M., Rousseau, G.X., Rozen, A., Sayad, E., van Schaik, L., Scharenbroch, B.C., Schirrmann, M., Schmidt, O., Schröder, B., Seeber, J., Shashkov, M.P., Singh, J., Smith, S.M., Steinwandter, M., Talavera, J.A., Trigo, D., Tsukamoto, J., de Valença, A.W., Vanek, S.J., Virto, I., Wackett, A.A., Warren, M.W., Wehr, N.H., Whalen, J.K., Wironen, M.B., Wolters, V., Zenkova, I.V., Zhang, W., Cameron, E.K., Eisenhauer, N.: Global distribution of earthworm diversity. Science **366**, 480–485 (2019). https://doi.org/10.1126/science.aax4851
4. van den Hoogen, J., Geisen, S., Routh, D., Ferris, H., Traunspurger, W., Wardle, D.A., de Goede, R.G.M., Adams, B., Ahmad, W., Andriuzzi, W.S., Bardgett, R.D., Bonkowski, M., Campos-Herrera, R., Cares, J.E., Caruso, T., Caixeta, L.D.B., Chen, X., Costa, S.R., Creamer, R., José Mauro da, D.M., Djibril, D., Miguel, E., Griffiths, B.S., Gutiérrez, C., Hohberg, K., Kalinkina, D., Kardo, P., Kergunteuil, A., Korthals, G., Krashevska, V., Kudrin, A.A., Li, Q., Liang, W., Magilton, M., Marais, M., Martín, J.A.R., Matveeva, E., Mayad, E.H., Mulder, C., Mullin. P., Neilson, R., Nguyen, T.A.D., Nielsen, U.N., Okada, H., Rius, J.E.P., Pan, K., Peneva, V/, Pellissier. L., da Silva, J.C.P., Pitteloud, C., Powers, T.O., Powers, K., Quist, C.W., Rasmann, S., Moreno, S.S., Scheu, S., Setälä, H., Sushchuk, A., Tiunov, A.V., Trap, J., van der Putten, W., Vestergård, M., Villenave, C., Waeyenberge, L., Wall, D.H., Wilschut, R., Wright, D.G., Yang, J., Crowther, T.W.: Soil nematode abundance and functional group composition at a global scale. Nature **572**, 194198 (2019). https://doi.org/10.1038/s41586-019-1418-6
5. Benckiser, G. (ed.): Fauna in Soil Ecosystems—Recycling Processes, Nutrient Fluxes, and Agricultural Production. Marcel Dekker, New York (1997a)
6. Benckiser, G.: Organic inputs and soil metabolism. In: Benckiser, G. (ed.) Fauna in Soil Ecosystems—Recycling Processes, Nutrient Fluxes, and Agricultural Production. Marcel Dekker, New York (1997b).
7. Benckiser, G., Schnell, S.: Biodiversity in Agricultural Production Systems. Taylor and Francis, Boca Raton (2007)
8. Díaz-Reviriego, I., Turnhout, E., Beck, S.: Participation and inclusiveness in the Intergovernmental Science–policy platform on biodiversity and ecosystem services. Nat. Sustain. **2**, 457–464 (2019). https://doi.org/10.1038/s41893-019-0290-6
9. Eisenhauer, N., Guerra, C.A.: Global maps of soil-dwelling nematode worms. Nature **572**, 187–188 (2019). https://doi.org/10.1038/d41586-019-02197-0
10. Midha, A., Schlosser, J., Hartmann, S.: Reciprocal interactions between nematodes and their microbial environments. Front. Cell Infect. Microbiol. **7**, 144 (2017). https://doi.org/10.3389/fcimb.2017.00144
11. Räty, M., Huhta, V.: Earthworms and pH affect communities of nematodes and enchytraeids in forest soil Biol Fertil Soils **38**, 52–58 (2003). https://doi.org/10.1007/s00374-003-0614-5

12. Riemann, F. The mouthless deep-sea nematode *Rhaptothyreus minor sp.* n. (Rhaptothyreidae, mermithoid nematodes of uncertain systematic position). Zool Scripta **22**, 341–346 (2005). https://doi.org/10.1111/j.1463-6409.1993.tb00363.x
13. Wang, X., Li, G.-H., Zou, C.-G., Ji, X.-L., Liu, T., Zhao, P.-J., Liang, M., Xu, J.-P., An, Z.-Q., Zheng, X., Qin, Y.-K., Tian, M.-Q., Xu, Y.-Y., Ma, Y.-C., Yu, Z.-F., Huang, X.-W., Liu, S.-Q., Niu, X.-M., Yang, J.-K., Huang, Y., Zhang, K.-Q. (2014) Bacteria can mobilize nematode-trapping fungi to kill nematodes. Nature Com. **5**, Article number: 5776 https://doi.org/10.1038/ncomms6776
14. Menta, C.: Soil fauna diversity—function, soil degradation biological indices, soil restoration. Open access peer-reviewed chapter (2012). https://doi.org/10.5772/51091
15. Blackmore, R.J.: Critical decline of earthworms from organic origins under intensive. Humic SOM-Depleting Agriculture Soil Syst. **2**(2), 33 (2018). https://doi.org/10.3390/soilsystems2020033
16. Whalen, J.K., Wironen, M.B., Wolters, V., Zenkova, I.V., Zhang, W., Cameron, E.K., Eisenhauer, N.: Global distribution of earthworm diversity. Science **366**, 480–485 (2019). https://doi.org/10.1126/science.aax4851
17. Bardgett, R.D., Usher, M.B., Hopkins, D.W. (eds.): Biological diversity and function in soils. Cambridge University (2005). Press ISBN-13 978–0–511–13511–8. eBook (EBL)
18. Briones, M.J.I.: The serendipitous value of soil fauna in ecosystem functioning: the unexplained explained. Front. Environ. Sci. (2018). https://doi.org/10.3389/fenvs.2018.00149
19. Brown, G.G.: How do earthworms affect microfloral and faunal community diversity? Plant Soil **170**, 209–231 (1995). https://doi.org/10.1007/BF02183068
20. Brussaard, L., De Ruiter, P.C., Brown, G.B.: Soil biodiversity for agricultural sustainability. Agric. Ecosys. Environ. **121**, 233–244 (2007). https://doi.org/10.1016/j.agee.2006.12.013
21. Decaens, T., Jimenez, J.J., Gioia, C., Measey, G.J., Lavelle, P.: The values of soil animals for conservation biology. Eur J Soil Biol **42**(Suppl. 1), S23–S38 (2006). https://doi.org/10.1016/j.ejsobi.2006.07.001
22. Furlong, M.A., Singleton, D.R., Coleman, D.C., Whitman, W.B.: Molecular and culture-based analyses of prokaryotic communities from an agricultural soil and the burrows and casts of the earthworm *Lumbricus rubellus*. Appl. Environ. Microbiol. **68**, 1265–1279 (2002). https://doi.org/10.1128/AEM.68.3.1265-1279.2002
23. Gardi, C., Jefferey, S.: Soil biodiversity. JRC Sci Techn. Rep. https://doi.org/10.2788/7831 (2009)
24. Jansa, J., Forczek, S.T., Rozmoš, M., Püschel, D., Bukovská, P., Hršelová, H.: Arbuscular mycorrhiza and soil organic nitrogen: network of players and interactions. Chem. Biol. Technol. Agric. **6**, 10 (2019). https://doi.org/10.1186/s40538-019-0147-2
25. Kopiejewska, W., Komosinski, K., Charkiewicz, J., Krupa, R.: Numbers and biomass of Lumbricidae on pastures treated with organic or mineral fertilisation within the area of Mazury landscape park in northeastern Poland. Polish J. Nat. Sci. **22**, 691–703 (2007)
26. Schneider, K.D., Thiessen Martens, J.R., Zvomuya, F., Reid, D.K., Fraser, T.D., Lynch, D.H., O'Halloran, I.P., Wilson, H.F.: Options for improved phosphorus cycling and use in agriculture at the field and regional Scales. J. Environ. Qual. **48**, 1247–1264 (2019). https://doi.org/10.2134/jeq2019.02.0070
27. Turbé, A., De Toni, A., Benito, P., Lavelle, P., Ruiz, N., Van der Putten, W.H., Labouze, E., Mudgal, S.: Soil biodiversity: functions, threats and tools for policy makers. Bio Intelligence Service, IRD, and NIOO. Report for European Commission (DG Environment) (2010). https://doi.org/10.2779/14571
28. Young, L.M., Crawford, J.W.: Interactions and selforganization in the soil-microbe complex. Science **304**, 1634–1637 (2004). https://doi.org/10.1126/science.1097394
29. Van der Heijden, M.A.G., Bardgett, R.D., Van Straalen, N.M.: The unseen majority: soil microbes as drivers of plant diversity and productivity in terrestrial ecosystems. Ecol. Lett. **11**, 296–310 (2007). https://doi.org/10.1111/j.1461-0248.2007.01139.x
30. Channarayappa, C., Biradar, D.P.: From Soil Basics, Management and Rhizosphere Engineering for Sustainable Agriculture. CRC Press, Taylor and Francis Group (2018). ISBN 9,781,138,486,928 –CAT# K349522

31. Bilalis, D., Sidiras, N., Vavoulidou, E., Konstantas, A.: Earthworm populations as affected by crop practices on clay loam soil in a Mediterranean climate. Acta Agric. Scan Section B Plant Soil Sci. **59**, 440–446 (2009). https://doi.org/10.1080/09064710802342327
32. Brito-Vega, H., Espinosa-Victoria, D.: Bacterial diversity in the digestive tract of earthworms (Oligochaeta). J. Biol. Sci. **9**, 192–199 (2009). https://doi.org/10.3923/jbs.2009.192.199
33. de Vries, F., Liiri, M., Bjørnlund, L., Bowker, M.A., Christensen, S., Setälä, H.M., Bardgett, R.D.: Land use alters the resistance and resilience of soil food webs to drought. Nature Clim. Change **2**, 276–280 (2012). https://doi.org/10.1038/nclimate1368
34. Whalley, W.R., Dumitru, H., Dexter, A.R.: Biological effects of soil compaction. Soil Tillage Res. **35**, 53–68 (1995). https://doi.org/10.1016/0167-1987(95)00473-6
35. Berg, M.P., Bengtsson, J.: Temporal and spatial variability in soil food web structure. Oikos **116**, 1789–1804 (2007). https://doi.org/10.1111/j.0030-1299.2007.15748.x
36. Carrillo, Y., Ball, B.A., Bradford, M.A., Jordan, C.F., Molina, M.: Soil fauna alter the effects of litter composition on nitrogen cycling in a mineral soil. Soil Biol Biochem **43**, 1440–1449 (2011). https://doi.org/10.1016/j.soilbio.2011.03.011
37. Hoppichler, J., Blab, A., Götz, B., Nowak, H., Oberleitner, I., Paar, M., Schwarzl, B., Zethner, G.: Biodiversity evaluation in the Alpes OECD-Fallstudie Forschungsbericht Nr. 48 BA für Bergbauernfragen (2002)
38. Willer, H., Yussefi-Menzler, M., Sorensen, N. (eds.).: The world of organic agriculture. statistics and emerging. Trends Int. Fed. Org. Agric. Movements (IFOAM), Bonn, Germany and Res. Inst. Org. Agri. (FiBL), Frick, Switzerland. ISBN IFOAM 978–3–934055–99–5 ISBN FiBL 978–3–03736–014–9
39. Feng, Y., Balkcom, K.S.: Nutrient cycling and soil biology in row crop systems under intensive tillage. In: Al-Kaisi, M.M., Lowery, B. (eds.) Health and Intensification of Agroecosytems, pp. 231–255 (2017). https://doi.org/10.1016/B978-0-12-805317-1.00011-7
40. Beylich, A., Oberholzer, H.R., Schrader, S., Höper, H., Wilke, B.M.: Evaluation of soil compaction effects on soil biota and soil biological processes in soils. Soil Tillage Res. **109**(2), 133–143 (2010). https://doi.org/10.1016/j.still.2010.05.010
41. Blakemore, R.J.: Ecology of earthworms under the "Haughley Experiment" of organic and conventional management regimes. Biol. Agric. Hortic. **18**, 141–159 (2000). https://doi.org/10.1080/01448765.2000.9754876
42. Moos, J.H., Schrader, S., Paulsen, H.M., Rahmann, G.: Occasional reduced tillage in organic farming can promote earthworm performance and resource efficiency. Appl. Soil Ecol. **103**, 22–30 (2016). https://doi.org/10.1016/j.apsoil.2016.01.017873
43. House, G.J., Parmelee, R.W.: Comparison of soil arthropods and earthworms from conventional and no-tillage agroecosystems. Soil Tillage Res. **5**, 351–360 (1985). https://doi.org/10.1016/S0167-1987(85)80003-9)
44. van Capelle, C., Schrader, S., Brunotte, J.: Tillage-induced changes in the functional diversity of soil biota—a review with a focus on German data. Eur. J. Soil. Biol. **50**, 165–181 (2012). https://doi.org/10.1016/j.ejsobi.2012.02.005
45. Roger-Estrade, J., Anger, C., Bertrand, M., Guy, R.: Tillage and soil ecology: partners for sustainable agriculture. Soil Tillage Res. **111**, 33–40 (2010). https://doi.org/10.1016/j.still.2010.08.010
46. Zunke, U., Perry, R.N.: Nematodes: harmful and beneficial organisms. In: Benckiser, G. (ed.) Fauna in soil ecosystems—recycling processes, nutrient fluxes, and agricultural production, pp. 85–133. Marcel Dekker, New York (1997)
47. Procheş, S., Syd, R.: The World's zoogeographical regions confirmed by cross-taxon analyses. Bioscience **62**, 260–270 (2012). https://doi.org/10.1525/bio.2012.62.3.7
48. Shipitalo, M.J.: Structure: Earthworms. Encyclopedia of Soil Science, 3rd edn. (2002). 10.1081/E-ESS3-1200537872212
49. Crotty, F.V., Fychan, R., Sanderson, R., Rhymes, J.R., Bourdin, F., Scullion, J., Marley, C.L.: Understanding the legacy effect of previous forage crop and tillage management on soil biology, after conversion to an arable crop rotation. Soil Biol. Biochem. **103**, 241–252 (2016). https://doi.org/10.1016/j.soilbio.2016.08.018

50. Frelich, L.E., Hale, C.M., Scheu, S., Holdsworth, A.R., Heneghan, L., Bohlen, P.J., Reich, P.B.: Earthworm invasion into previously earthworm-free temperate and boreal forests. Biol. Invas. **8**, 1235–1245 (2006). https://doi.org/10.1007/s10530-006-9019-3866
51. Radford, B.J., Wilson-Rummenie, A.C., Simpson, G.B., Bell, K.L., Ferguson, M.A.: Compacted soil affects soil macrofauna populations in a semi-arid environment in central Queensland. Soil Biol. Biochem. **33**, 1869–1872 (2001). https://doi.org/10.1016/S0038-0718817(01)00104-3
52. Wironen, M., Moore, T.R.: Exotic earthworm invasion increases soil carbon and nitrogen in an old-growth forest in southern Quebec. Can. J. For. Res. **36**, 845–854 (2006). https://doi.org/10.1139/X06-016
53. Jernigan, A.B., Wickings, K., Mohler, C.L., Caldwell, B.A., Pelzer, C.J., Wayman, S., Ryan, M.R.: Legacy effects of contrasting organic grain cropping systems on soil health indicators, soil invertebrates, weeds, and crop yield. Agric. Sys. **177**, 102719. https://doi.org/10.1016/j.agsy.2019.102719 (2020)
54. Enkegaard, A., Br Ødsgaard, H.: *Lasioseius fimetorum*: a soil-dwelling predator of glasshouse pests? Biocontrol **45**, 285 (2000). https://doi.org/10.1023/A:1009981920402
55. Amador, J.A., Avizinis, E.J.: Response of nitrous oxide flux to addition of anecic earthworms to an agricultural field. J. Soil. Sci. **3**, 100–106 (2013). https://doi.org/10.4236/ojss.2013.32011
56. Ashworth, A.J., Allen, F.L., Tyler, D.D., Pote, D.H., Shipitalo, M.J.: Earthworm populations are affected from long-term crop sequences and bio-covers under no-tillage. Pedobiologia **60**, 27–33 (2017). https://doi.org/10.1016/j.pedobi.2017.01.001844137
57. Blakemore, R.J.: Miscellaneous earthworm types in the Natural History Museum, London (Annelida: Oligochaeta: Megadrilacea: Eudrilidae, Lumbricidae, Megascolecidae, Moniligastridae, Octochaetidae). Opuscula Zool. **45**, 199–155 (2014). urn:lsid:zoobank.org:pub:9451138A-8E68–43A3-B580-D5C1DA612CE7
58. Friedl, J., Scheer, C., Rowlings, D.W., Deltedesco, E., Gorfer, M., De Rosa, D., Grace, P.R., Müller, C., Keiblinger, K.M.: Effect of the nitrification inhibitor 3,4-dimethylpyrazole phosphate (DMPP) on N-turnover, the N_2O reductase-gene *nosZ* and N_2O:N_2 partitioning from agricultural soils. Sci. Rep. (in press, 2020)
59. Pfeiffer, M., Chimedregzen, L., Kaman Ulykpan, K.: Community organization and species richness of ants (Hymenoptera/Formicidae) in Mongolia along an ecological gradient from steppe to Gobi desert. J. Biogeogr. **30**, 1921–1935 (2003). https://doi.org/10.1046/j.0305-0270.2003.00977.x
60. Benckiser, G.: Ants and sustainable agriculture. A review. Sustain. Dev. **30**, 191–199 (2010). https://doi.org/10.1051/agro/2009026
61. Kwon, T.-S.: Ants foraging on grasses in South Korea: high diversity in Jeju Island and negative correlation with aphids. J Asia-Pacific Biodiv. **10**, 465–471 (2017). https://doi.org/10.1016/j.japb.2017.06.002
62. Blanchart, E., Lavelle, P., Braudeau, E., Le Bissonnais, Y., Valentin, C.: Regulation of soil structure by geophagous earthworm activities in humid savannas of Côte d'Ivoire. Soil Biol. Biochem. **29**, 431–439 (1997). https://doi.org/10.1016/S0038-0717(96)00042-9
63. Benckiser, G., Stuart, S., Bamforth, S.S.: Role of pathogens, signal recalcitrance, and organisms shifting for ecosystem recuperation. A review. Agron. Sustain. Dev. **31**, 205–215 (2011). https://doi.org/10.1051/agro/20100241
64. Benckiser, G.: Plastics, micro- and nanomaterials, and virus-soil microbe-plant interactions in the environment. In: Prasad, R. (ed.) Plant Nanobionics, Nanotechnology in the Life Sciences. Springer Nature, Switzerland, p. 83 (2019). https://doi.org/10.1007/978-3-030-12496-0_4
65. Djigal, D., Baudoin, E., Philippot, L., Brauman, A., Villenave, C.: Shifts in size, genetic structure and activity of the soil denitrifier community by nematode grazing. Eur. J. Soil Biol. **46**, 112–118 (2010). https://doi.org/10.1016/j.ejsobi.2009.12.001
66. Geisen, S., Mitchell, E.A.D., Wilkinson, D.M., Ad, S., Bonkowski, M., Brown, M.W., Fiore-Donno, A.M., Heger, T.J., Jassey, V.E.J., Krashevska, V., Lahr, D.J.G., Marcisz, K., Mulot, M., Payne, R, Singer, D., Anderson, O.R., Charma, D.J., Ekelund, F., Griffiths, B.S., Ronn, R.,

Smirnov, A., Bass, D., Balbahri, L., Berney, C., Blandenier, Q., Chatzinotas, A., Clarholm, M., Dunthorn, M., Feest, A., Fernández, L.D., Foissner, W., Fournier, B., Gentekaki, E., Hájek, M, Helder, J, Jousset, A., Koller, R., Kumar, S., La Terza, A., Lamentowicz, M., Mazei, Y., Santos, S.S. Seppey, C.V.W., Spiegel, F.W., Walochnik, J., Winding, A., Lara, E.: Soil protistology rebooted: 30 fundamental questions to start with. Soil Biol. Biochem. **111**, 94–103 (2017). https://doi.org/10.1016/j.soilbio.2017.04.001

67. Ingham, R.E., Trofymow, J.A., Ingham, E.R., Coleman, D.C.: Interactions of bacteria, fungi, and their nematode grazers: effects on nutrient cycling and plant growth. Ecol. Monogr. **55**, 119–140 (1985). https://doi.org/10.2307/1942528
68. Michalsky, R., Pfromm, P.H., Steinfeld, A.: Rational design of metal nitride redox materials for solar-driven ammonia synthesis. Interface Focus **5**, 20140084 (2015). https://doi.org/10.1098/rsfs.2014.0084
69. Schmidt, A.C., Fraser, L.H., Carlyle, C.N., Bassett, E.R.L.: Does cattle grazing affect ant abundance and diversity in temperate Grasslands? Rangeland Ecol. Manag. **65**, 292–298 (2012). https://doi.org/10.2111/REM-D-11-00100.1
70. Schorpp, Q., Schrader, S.: Dynamic of nematode communities in energy plant cropping systems. Eur. J. Soil Biol. **78**, 92–101 (2017). https://doi.org/10.1016/j.ejsobi.2016.12.002
71. Bamforth, S.S.: Diversity of protozoa. In: Benckiser, G., Schnell, S. (eds.) Biodiversity in Agricultural Production Systems. Taylor and Francis, Boca Raton, FL (2007)
72. Jäckle, O., Seah, B.K., Tietjen, M., Leisch, N., Liebeke, M., Kleiner, M., Berg, J., Gruber-Vodicka, H.R.: Chemosynthetic symbiont with a drastically reduced genome serves as primary energy storage in the marine flatworm *Paracatenula*. PNAS **116**, 8505–8514 (2019). https://doi.org/10.1073/pnas.1818995116
73. Schulenburg, H., Félix, M.A.: The Natural biotic environment of *Caenorhabditis elegans*. Genetics **206**, 55–86 (2017). https://doi.org/10.1534/genetics.116.195511
74. Cameron, T.C., Plaistow, S., Mugabo, M., Piertney, S.B., Benton, T.G.: Eco-evolutionary dynamics: experiments in a model system. Adv Ecol. Res **50**, 172–201 (2014). https://doi.org/10.1016/B978-0-12-801374-8.00005-0
75. Chen, C.-G., Chen, T., Hua, B.-Z., Wan, T.-R.: Structure and functions of the ventral tube of the clover springtail *Sminthurus viridis* (Collembola: Sminthuridae). Sci. Rep. **9**, 897 (2019). https://doi.org/10.1038/s41598-018-37354-4
76. Scheller, U.: Taxonomic and distributional notes on pauropods from the United States (Myriapoda, Pauropoda: Pauropodidae, Eurypauropodidae) In: Insect Systematics & Evolution, vol. 16, pp. 237–257 (1985). https://doi.org/10.1163/187631285X00135
77. Tipping, C., Allen, R.T.: Description of two new species of Eosentomon from the Ouachita Mountains of Arkansas with a key to the species with the 6/4 setal pattern on sterna IX/X (Protura: Eosentomidae). J. New York Entomol. Soc. **103**, 287–301 (1996)
78. Brusca, R.C., Wilson, G.F.: A phylogenetic analysis of the Isopoda with some classificatory recommendations. Mem. Queensland Museum **31**, 143–204 (1991). Brisbane. ISSN 0079-8835
79. Halaj, J., Cady, A.B.: Diet composition and significance of earthworms as food of harvestmen (Arachnida: Opiliones). Am. Midland Nat. **143**, 487–491 (2000). http://dx.doi.org/10.1674/0003-0031(2000)143[0487:DCASOE]2.0.CO;2
80. Heiling, A.M., Herberstein, M.E.: The web of *Nuctenea sclopetaria* (Araneae, Araneidae): relationship between body size and web design. J. Arachnol. **26**, 91–96 (1998). https://doi.org/10.2307/3544406
81. Stöckli, E.: Literature-based survey on the Swiss fauna of Chilopoda. Soil Organ. **81**, 647–669 (2009)
82. Tropea, G., Fet, V., Parmakeli, A., Kotsakioski, P., Stahlil, I.: Redescription of *Euscorpius tauricus* (C.L. Koch, 1837), with the description of two new related species from Greece (Scorpiones: Euscorpiidae). Ecol. Mont. **7**, 614–638 (2017). www.biotaxa.org/em
83. Haynes, R.J.: Chapter two—nature of the belowground ecosystem and its development during pedogenesis. Adv. Agron. **127**, 43–109 (2014). https://doi.org/10.1016/B978-0-12-800131-8.00002-9

84. Wüst, P.K., Horn, M.A., Henderson, G., Janssen, P.H., Rehm, B., Drake, H.L.: Gut-associated denitrification and in vivo emission of nitrous oxide by the earthworm families *Megascolecidae* and *Lumbricidae* in New Zealand. Appl. Environm Microbiol. **75**, 3430–3436 (2009). https://doi.org/10.1128/AEM.00304-09
85. Judas, M.: The species-area relationship of European *Lumbricidae* (Annelida, Oligochaeta). Oecologia **76**, 579–587 (1988). https://doi.org/10.1007/BF00397874868
86. Omodeo, P., Rota, E.: Earthworms of Turkey. Italian J. Zool. **56**, 167–198 (1989). https://doi.org/10.1080/11250008909355639
87. Lakhani, K.H., Satchell, J.E.: Production by *Lumbricus terrestris* L. J. Anim. Ecol. **39**, 473–492 (1970). https://doi.org/10.2307/2983870
88. Benckiser, G.: Eusocial ant nest management, template for land development. WSEAS Trans. Environ. Dev. **13**, 204–215 (2017). E-ISSN: 2224–3496
89. Jiménez-Chacón, A., Homet, P., Matías, L., Gómez-Aparicio, L., Godoy, O.: Fine scale determinants of soil litter fauna on a Mediterraneasn mixed oak forest invaded by the exotic soil-borne pathogen *Phytophthora cinnamomi*. Forests **9**, 218 (2018). https://doi.org/10.3390/f9040218
90. Lee, K.E.: Earthworms: Their Ecology and Relationships with Land Use. Academic Press, Sydney (1985)
91. Lobry de Bruyn, L.A.: The status of soil macrofauna as indicators of soil health to monitor the sustainability of Australian agricultural soils. Ecol. Econom. **23**, 167–178 (1997). https://doi.org/10.1016/S0921-8009(97)00052-9
92. Brinkhurst, R.O., Nemec, A.F.L.: A comparison of phenetic and phylogenetic methods applied to the systematics of Oligochaeta. In: Brinkhurst, R.O., Diaz, R.J. (eds.) Aquatic Oligochaeta. Developments in Hydrobiology, p. 40. Springer, Dordrecht (1987). https://doi.org/10.1007/978-94-009-3091-9_7
93. Buchkowski, R.W., Shaw, A.N., Sihi, D., Smith, G.R., Keiser, A.D.: Constraining carcon and nitrogen flows in soil with ecological stoichiometry. In: Evans-White, M., Elser, J.J. (eds.) Emerging Frontiers in Ecological Stoichiometry Frontiers in Ecology and Evolution, Hypothesis and Theory, pp 362–394 (2019). 10.3383.2019.00382
94. Jamieson, B.M.G.: Phylogenetic and phonetic systematics of the opisthoporous Oligochaeta (Annelida: Clitellata). Evol. Theory **3**, 195–233 (1978)
95. Morris, S.C., Pickerick, R.K., Harland, T.L.: A possible annelid from the Trenton limestone (Ordovician) of Quebec, with a review of fossil oligochaets and other annulate worms. Can. J. Earth Sci. **19**, 2150–2157 (1982). https://doi.org/10.1139/e82-189
96. Osler, G.H.R., Sommerkorn, M.: Toward a complete soil C and N cycle: incorporating the soil fauna. Ecology **88**, 1611–1621 (2007). https://doi.org/10.1890/06-1357.1
97. Mısırlıoğlu, I.M., Tsekova, R., Stojanovic, M.: On the presence of *Lumbricus terrestris* Linnaeus 1758 (Oligochaeta, Lumbricidae) on the Balkan Peninsula: some aspects of ecology and distribution. Turkish J. Zool. **40**, 438–444 (2016). https://doi.org/10.3906/zoo-1509-12
98. Plaas, E., Meyer-Wolfarth, F., Banse, M., Bengtsson, J., Bergmann, H., Faber, J., Potthoff, M., Runge, T., Schrader, S., Taylor, A.: Towards valuation of biodiversity in agricultural soils: A case for earthworms. Ecol. Econ. **159**, 291–300 (2019). https://doi.org/10.1016/j.ecolecon.2019.02.003
99. Römbke, J., Dorow, W.H.O., Jänsch, S.: Distribution and diversity of earthworms (Lumbricidae) in Hesse (Central Germany): current knowledge. Soil Organ. **90**, 171–185 (2018). https://doi.org/10.25674/bmwx-qs58
100. Anderson, N.C.: Nitrogen turnover by earthworms in arable plots treated with farmyard manure and slurry. In: Satchell, J.E. (ed.) Earthworm Ecology from Darwin to Vermiculture (Chap. 11), pp. 139–150. Capman and Hall, London (1983)
101. Edwards, C.A., Lofty, J.R.: Effects of agriculture on earthworm populations. In: Biology of Earthworms, pp. 203–221. Springer, Boston (1977). https://doi.org/10.1007/978-1-4613-3382-1_9
102. Edwards, C.A., Bohlen, P.J.: Biology and Ecology of Earthworms, 3rd edn. Chapman and Hall, London (1996)

103. Satchell, J.E.: Earthworm microbiology. In: Satchell, J.E. (eds.) Earthworm Ecology, pp. 351–364. Springer, Dordrecht (1983). shttps://doi.org/10.1007/978-94-009-5965-1_31
104. Blouin, M., Hodson, M.E., Delgado, E.A., Baker, G., Blouin, M., Brussaard, L., Butt, K.R., Dai, J., Dendooven, L., Pérès, G., Tondoh, J.E., Cluzeau, D., Brun, J.J.: A review of earthworm impact on soil function and ecosystem services. Eur. J. Soil Sci. **64**, 161–182 (2013). https://doi.org/10.1111/ejss.12025
105. Fründ, C., Butt, K., Capowiez, Y., Eisenhauer, N., Emmerling, C., Ernst, G., Potthoff, M., Schädler, M., Schrader, S.: Using earthworms as model organisms in the laboratory: recommendations for experimental implementations. Pedobiologia **53**, 119–125 (2010). https://doi.org/10.1016/j.pedobi.2009.07.002
106. Chan, K.Y.: An overview of some tillage impacts on earthworm population abundance and diversity—implications for functioning in soils. Soil Till. Res. **57**, 179–191 (2001). https://doi.org/10.1016/S0167-1987(00)00173-2
107. Fierer, N., Strickland, M.S., Liptzin, D., Bradford, M.A., Cleveland, C.C.: Global patterns in belowground communities. Ecol Lett **12**, 1238–1249 (2009). https://doi.org/10.1111/j.1461-0248.2009.01360.x
108. Casagrande, M., Peigné, J., Payet, V., Mäder, P., Sans, F.X., Blanco-Moreno, J.M., Antichi, D., Bàrberi, P., Beeckman, A., Bigongiali, F., Cooper, J., Dierauer, H., Gascoyne, K., Grosse, M., Heß, J., Kranzler, A., Luik, A., Peetsmann, E., Surböck, A., Willekens, K., David, C.: Organic farmers' motivations and challenges for adopting conservation agriculture in Europe Organic. Agriculture **6**, 281–295 (2016). https://doi.org/10.1007/s13165-015-0136-0
109. Kuntz, M., Berner, A., Gattinger, A., Scholberg, J.M., Mäder, P., Pfiffner, L.: Influence of reduced tillage on earthworm and microbial communities under organic arable farming. Pedobiologia 56:251–260 (2013). https://doi.org/10.1016/j.pedobi.2013.08.005
110. Asshoff, R., Scheu, S., Eisenhauer, N.: Different earthworm ecological groups interactively impact seedling establishment. Eur. J. Soil Biol. **46**, 330–334 (2010). https://doi.org/10.1016/j.ejsobi.2010.06.005
111. Curry, J.P., Schmidt, O.: The feeding ecology of earthworms—A review. Pedobiologia **50**, 463–477 (2007). https://doi.org/10.1016/j.pedobi.2006.09.001
112. Pavlícek, T., Csuzdi, C., Nevo, E.: Species richness and zoogeographic affinities of earthworms in the Levant. In: The 7th International Symposium on Earthworm Ecology Cardiff Wales 2002. Pedobiologia **47**, 452–457 (2003). https://doi.org/10.1078/0031-4056-00212
113. Richard, B., Decaens, T., Rougerrie, R., James, S.W., Porco, D., Herbert, P.D.N.: Re-integrating earthworm juveniles into soil biodiversity studies: species identification through DNA barcoding. Mol Ecol Resour. **10**, 606–614 (2010). https://doi.org/10.1111/j.1755-0998.2009.02822.x
114. Tracic, T., Valkowski, H., Stojanovic, M.: Endemic earthworms (Oligochaeta: Lumbricidae) of the Balkan Peninsula: a review. Proc. Zool. Soc. London Ser. B **111**, 197–236 (2016)
115. Ababsa, N., Laiche, A., Djabbar, F.: Abundance and species richness of lombric macrofauna in a semi-arid forest ecosystem. J. Fundam. Appl. Sci. **9**, 1634–1643 (2017)
116. Lemtiri, A., Colinet, G., Alabi, T., Cluzeau, D., Zirbes, L., Haubruge, É., Francis, F.: Focus on: impacts of earthworms on soil components and dynamics. A review Biotechnol. Agron. Soc. Environ. **18**, 121–133 (2014)
117. Sims, R.W., Gerar, B.M.: A Synopsis of the Earthworms. Linnean Society of London, Pitman Press (1985). ISBN 90 04 07582 8, 168 p
118. Van Praagh, B.: The biology and conservation of the Giant Gippsland Earthworm *Megascolides australis* McCoy, 1878. Soil Biol. Biochem. **24**, 1363–1367 (1992). https://doi.org/10.1016/0038-0717(92)90119-I
119. Lynn, H., Booth, L.H., O'Halloran, K.: A comparison of biomarker responses in the earthworm *Aporrectodea caliginosa* to the organophosphorus insecticides diazinon and chlorpyrifos. Environ. Toxicol. **20**, 2494–2502 (2001). https://doi.org/10.1002/etc.5620201115
120. Wang, Y., Cang, T., Zhao, X., Yu, R., Chen, L., Wu, C., Qiang, W.: Comparative acute toxicity of twenty-four insecticides to earthworm. Eisenia Fetida Ecotoxicol. Environment. Saf. **79**, 122–128 (2012). https://doi.org/10.1016/j.ecoenv.2011.12.016

121. Reddell, P., Spain, A.V.: Earthworms as vectors of viable propagules of mycorrhizal fungi. Soil Biol. Biochem. **23**, 767–774 (1991). https://doi.org/10.1016/0038-0717(91)90147-C
122. Fierer, N.: Earthworms' place on Earth. Science **366**, 425–426 (2019). https://doi.org/10.1126/science.aaz5670
123. Lu, Y., Nakicenovic, N., Visbeck, M., Stevance, A.S.: Five priorities for the UN Sustainable Development Goals. Nature **520**, 432–433 (2015). https://doi.org/10.1038/520432a
124. Brown, G.G., Barois, I., Lavelle, P.: Regulation of soil organic matter dynamics and microbial activity in the drilosphere and the role of interactions with other edaphic functional domains. Eur. J. Soil Biol. **36**, 177–198 (2000). https://doi.org/10.1016/S1164-5563(00)01062-1
125. Pauli, N., Abbott, L.K., Negrete-Yankelevich, S., Andrés, P.: Farmers' knowledge and use of soil fauna in agriculture: a worldwide review. Ecol. Soc. **21**, 19 (2016). https://dx.doi.org/10.5751/ES-08597-210319
126. Vavoulidou E, Dimirkou A, Papadopoulos P, Avramides EJ, Dimitris, A (2002) A comparative study for the control of organic agriculture in a region of Greece. In: 17 WCSS Symposium, vol. 57, pp. 1–14, 737 p
127. Vavoulidou, E., Coors, A., Dózsa-Farkas, K., Römbke, J.: Influence of farming practice, crop type and soil properties on the abundance of Enchytraeidae (Oligochaeta) in Greek agricultural soils. Soil Organ. **81**, 197–212 (2009)
128. Solomon, A.D., Sfougaris, A.I., Vavoulidou, E.M., Csuzdi, C.: Species richness and density of earthworms in relation to soil factors in olive orchard production systems in central Greece Com Soil Sci. Plant. Anal. **44**, 301–311 (2013). https://doi.org/10.1080/00103624.2013.741904
129. Černosvitov, L.: Mission Scientifique de l'Omo. Oligochaeta. Mém. Mus. natu. Hist. nat. Paris **4**(38):253–318 (1938)
130. Savigny (1826) Eisenia fetida (Savigny, 1826):
131. Kinberg: (1867) Johan Gustaf Hjalmar Kinberg (1866): Annulata nova. Öfversigt af Kongliga Vetenskaps-Akademiens Förhandlingar 23 (4), S. 97–103, hier 97 (Amynthas n.) und 101 (Amyntas [sic] n.). https://de.wikipedia.org/wiki/Johan_Gustaf_Hjalmar_Kinberg
132. Szederjesi, T., Vavoulidou, E., Chalkia, C., Danyi, L., Csudi, C.: An annotated checklist of earthworms of Greece (Clitellata: Megadrili). Zootaxa **4272**, 57–82 (2017). https://doi.org/10.11646/zootaxa.4272.1.3ss
133. Parle, J.N.: A microbiological study of earthworm casts. J. Gen. Micro Mol. **31**, 13–22 (1963)
134. Pavlíček, T., Csuzdi, C.: Clitellata: Oligochaeta: earthworms (Chap. 23). In: Sparrow, D.J., Eddie, J. (eds.) An Introduction to the Wildlife of Cyprus. Terra Cypria, Limassol, pp. 587–599 (2016)
135. Szederjesi, T., Csuzdi, Cs.: New and little known earthworm species form Greece (Oligochaeta: Lumbricidae, Acanthodrilidae). Zootaxa **3304**, 25–42 (2012)
136. Edwards, C.A., Lofty, J.R.: Nitrogenous fertilizers and earthworm populations in agricultural soils. Soil Biol. Biochem. **14**, 515–521 (1982). https://doi.org/10.1016/0038-0717(82)90112-2
137. Botinelly, N., Henry-des-Tureaux, T., Hallaire, V., Mathieu, J., Benard, Y., Duc, T., Jouquet, P.N.: Earthworms accelerate soil porosity turnover under watering conditions. Geoderma **156**, 43–47 (2010). https://doi.org/10.1016/j.geoderma.2010.01.006
138. Jouquet, P., Huchet, G., Bottinelli, N., Thu, T.D.: What are the limits of the drilosphere? An incubation experiment using Metaphire posthuma Pedobiol **54**(Supplement), S113–S117 (2011). https://doi.org/10.1016/j.pedobi.2011.07.013
139. Vavoulidou, E., Avramides, E.J., Papadopoulos, P., Dimirkou, A.: Copper content in agricultural soils related to cropping systems in different regions of Greece. Commun. Soil Sci. Plant Anal. **36**, 759–773 (2005). https://doi.org/10.1081/CSS-200043367
140. Streit, B.: (1984) Effects of high copper concentrations on soil invertebrates (earthworms and oribatid mites): : Experimental results and a model. Oecologia **64**(3), 381–388 (1984). https://doi.org/10.1007/BF00379137
141. Boyle, K, Curry, J.P., Farrell, E.: Influence of earthworms on soil properties and grass production in reclaimed cutover peat. Biol. Fertil. Soils **25**, 20–26 (1997). https://doi.org/10.1007/s003740050274

142. Curry, J.P., Doherty, P., Purvis, G., Schmidt, O.: Relationships between earthworm populations and management intensity in cattle-grazed pastures in Ireland. Appl Soil Ecol **39**, 58–64 (2008). https://doi.org/10.1016/j.apsoil.2007.11.005
143. Robinson, C.H., Ineson, P., Piearce, T.G., Rowland, A.P.: Nitrogen mobilization by earthworms in Limed peat soils under *Picea sitchensis*. J Appl Ecol **29**, 226–237 (1992). https://doi.org/10.2307/2404365
144. Schmidt, O., Curry, J.P.: Population dynamics of earthworms (Lumbricidae) and their role in nitrogen turnover in wheat and wheatclover cropping systems. Pedobiologia **45**, 174–187 (2001). https://doi.org/10.1078/0031-4056-00078
145. Vavoulidou, E., Avramides, E.J., Dimirkou, A., Papadopoulos, P.: Influence of different cultivation practices on the properties of volcanic soils on Santorini Island. Greece Commun. Soil Sci. Plant. Anal. **37**, 2857–2866 (2006). https://doi.org/10.1080/00103620600832837
146. Ostermann, O.P.: The need for management of nature conservation sites designated under Natura 2000. J. Appl. Ecol. **35**, 968–973 (2008). https://doi.org/10.1111/j.1365-2664.1998.tb00016.x
147. Phelps, J., Webb, E.L., Bickford, D., Nijman, V., Sodhi, N.S.: Boosting CITES Sci. **330**, 1752–1753 (2020). https://doi.org/10.1126/science.1195558
148. Benirschke, K.: Frozen Zoo®. San Diego Zoobiol. **3**, 325–328 (1984). https://doi.org/10.1002/zoo.1430030405
149. Vaas, L.A.I., Sikorski,J., Hofner, B., Fiebig, A., Buddruhs, N., Klenk, H.-P.,Göker. M.: opm: an R package for analysing OmniLog® phenotype microarray data. Bioinformatics **29**, 1823–1824 (2013). https://doi.org/10.1093/bioinformatics/btt291
150. Westengen, O.T., Jeppson, S., Guarino, L.: Global ex-situ crop diversity conservation and the Svalbard global seed vault: assessing the current status. PLoS One **8**, e64146 (2013). shttps://doi.org/10.1371/journal.pone.0064146
151. Ponge, J.F., Chevalier, R., Loussot, P.: Humus Index. SSSAJ **66**, 1996–2001 (2002). https://doi.org/10.2136/sssaj2002.1996
152. Bonkowski, M., Griffiths, B.S., Ritz, K.: Food preferences of earthworms for soil fungi. Pedobiologia **44**, 666–676 (2000). https://doi.org/10.1078/S0031-4056(04)70080-3
153. McGonigle, T.P.: Effects of animals grazing on fungi. In: Kubicek, C., Druzhinina, I. (eds.) Environmental and Microbial Relationships. The Mycota, vol. 4. Springer, Berlin (2007). https://doi.org/10.1007/978-3-540-71840-6_12
154. Filser, J., Faber, J.H., Tiunov, A.V., Brussaard, L., Frouz, J., De Deyn, G., Uvarov, A.V., Berg, M.P., Lavelle, P., Loreau, M., Wall, D.H., Querner, P., Eijsackers, H., Jiménez, J.J.: Soil fauna: key to new carbon models. Soil **2**(565–582), 2016 (2016). https://doi.org/10.5194/soil-2-565-2016
155. van der Putten, W.H.: Trends in soil biodiversity research and applications. In: Abstract from The First GSBI Conference—Assessing Soil Biodiversity and its Role for Ecosystem Services, Dijon, France (2014)
156. Díaz, S., Demissew, S., Carabias, J., Joly, C., Lonsdale, M., Ash, N., Larigauderie, A., Adhikari, J.R., Arico, S., Báldi, A., Bartuska, A., Baste, I.A., Bilgin, A., Brondizio, E., Chan, K.M., Figueroa, V.E., Duraiappah, A., Fischer, M., Diana, Z.: The IPBES conceptual framework—connecting nature and people. Curr. Opin. Environ. Sustain. **14**, 1–16. https://doi.org/10.1016/j.cosust.2014.11.002

Synthetic Biology and the Possibilities in Achieving a Plant Demand and Soil Buffer Capacity Adapted Nitrogen (N) Recycling

Lena Schorr, Janina Schoen, and Gero Benckiser

Abstract The world demand for nitrogen (N) fertilizer increases since Haber and Bosch's invention of technical nitrogen fixation (TNF) with an average growth of 1.9% per annum or an amount of 117.116 tons N in 2019 (FAO in World Fertilizer Trends and Outlook to 2020, 2017, [34]). Landscapes not only receive TNF, but also the wastes from industrial livestock farming, an increasing industrialisation, a growing population and from overloading wastewater treatment plants by the production of sewage sludge. Thus, the administration is forced to handle and for instance the German capital Berlin hired 14.364 ha land in 1913 and installed a sewage farm for spreading Berlin's wastewater. In 1926, the amount of heavy-metal polluted wastewater produced by Berlin's industry was 7.3%. In 1988, still 34% of all accruing wastewater was spread on the fields [61]. From Berlin's nutrient over satiated sewage farm soil, alarming amounts of nitrate flew into nearby glacial valley wells, surface waters and riverbank filtrates, which are used as drinking water sources and thus, in 1988 the wastewater spreading on the sewage farm stopped. Synthetic biology is a promising approach to reduce soil N buffer capacity over satiating concentrations. The possibilities of synthetic biology techniques in the focus of this chapter are discussed towards curing N recycling related impacts on health, soil management and wasterwater treatment plants.

Keywords Nitrogen pollution · Nitrification · Dinitrification · Synthetic biology · Soil and wasterwater management · Chassis organisms · Nitrogen recycling · Anammox

L. Schorr
Integrated Immunology, Friedrich-Alexander-Universität Erlangen-Nürnberg, Erlangen, Germany
e-mail: lena.schorr@fau.de

J. Schoen (✉)
Cell- and Molecular Biology, Friedrich-Alexander-Universität Erlangen-Nürnberg, Erlangen, Germany
e-mail: ja.schoen@gmx.de

G. Benckiser
Institute of Applied Microbiology, Justus-Liebig University, Heinrich Buff-Ring 26-32, Giessen, Germany

G. Benckiser (ed.), *Soil and Recycling Management in the Anthropocene Era*, Environmental Science and Engineering,
https://doi.org/10.1007/978-3-030-51886-8_5

1 Nitrogen Pollution and Health Effects

Wastewater disposal and plant demand surpassing N fertilization in form of urea (CH_4N_2O), ammonium (NH_4^+) and nitrate (NO_3^-) pollute surface and groundwater, landscapes and the atmosphere. Thus, organisms, evolutionary adapted to N shortage, are forced to come along with N overloads. Microbes have the potential to oxidize NH_4^+ to NO_3^- and to respire NO_3^- to NO_2^-, further to climate change contributing NO_x and nitrous oxide (N_2O) greenhouse gases, to N_2 and even to ammonia (NH_3) [7, 69, 90]. Water soluble nitrate and toxic NO_2^-, if available or applied in plant demand surpassing amounts, pollute surface and ground waters. Nitrate sources are agricultural and non-agricultural spreaders, which produce increasing amounts of wastewater that reaches wastewater treatment plants (WWTPs) and leads to huge amounts of sewage sludge [105, 111]. For preventing high nitrate levels in the drinking water, the world health organisation (WHO) has issued that groundwater NO_3^- should not exceed 50 mg/l, because nitrate overloads lead denitrification related (Fig. 1) to an increased emission of the greenhouse gas N_2O, to eutrophication, hypoxia, and

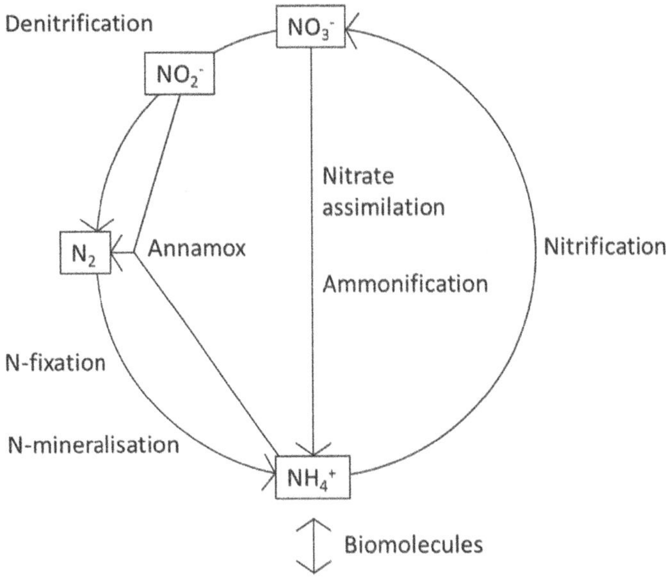

Fig. 1 Nitrification and denitrification are central parts of the nitrogen cycle and various involved bacteria, archaea, and fungi species return the fixed nitrogen, preferred as N_2 back to the atmosphere, after transforming nitrate via assimilation, ammonification, the anammox process or denitrification. During nitrification, ammonia or ammonium is oxidized to nitrate via the intermediate NO_2^-. Denitrification describes the conversion of NO_3^- to N_2 via the intermediates NO_2^-, NO and N_2O. The cycle is closed by N-fixation and mineralisation that leads to the formation of NH_4^+. The whole nitrogen cycle is thereby completed by nitrate assimilation and ammonification, as well as the anaerobic ammonium oxidation (anammox), whereby NH_4^+ and NO_2^- are converted into N_2 (after [102])

to changes in landscape habitats [41, 97, 116, 117]. Further, the consumption of drinking water with 50 mg/l surpassing nitrate concentrations may cause methemoglobinemia, cancer, birth defects and health restrictions in general. Within the gastrointestinal tract, bacteria as the *Lactobacillus* species may reduce nitrate to nitrite that can react with the ferrous heme group of deoxygenated haemoglobin and result in methemoglobinemia with hypoxia as typical outcome [6, 40, 114]. In particular children, not yet possessing the methaemoglobin reductase in contrast to adults, are suffering ("blue baby syndrome"). High nitrate concentrations consumed by mothers may affect the neural tube and the central nervous system of the unborn child. The formation of genotoxic, carcinogenic nitroso-compounds from NO_2^- overloads may affect also the health of adults through colorectal cancer, thyroid disease, high blood pressure, and infections in the respiratory tract [6, 19, 27, 43, 80, 83, 89, 96, 98, 115]. Such health risks afford to understand the processes involved in nitrate reduction.

2 The N-Recycling Processes Nitrification and Denitrification

Global nitrification and denitrification depend on biological and technical N_2 fixation, the introduction of ammonia in the N cycle, the oxidation of NH_3 to nitrate under energy gain and biomass formation, and the reduction of NO_3^- to NO_2^-, NO, N_2O, N_2, or even to NH_3 also out of energy gain and biomass formation reasons (Fig. 1). The intermediate products NO_3^-, NO_2^-, NO, N_2O of nitrification and denitrification as well as NH_3 co-shape eutrophication, the atmosphere and the climate [14, 17, 11, 15].

A broad spectrum of nitrifying bacteria and archaea, such as *Nitrosomonas europaea* that oxidizes ammonia to the intermediates hydroxylamine, nitrite, and finally to nitrate, pursue an autotroph living style and are present in soils and WWTPs. With electrons (e^-) and protons (H^+), elevated to a higher energetic level during photosynthesis and stored in NH_3, nitrifying bacteria and archaea gain energy by oxidizing NH_3 to nitrate and cell biomass by reducing CO_2 with the in NH_3 stored e^- and H^+ (Fig. 1; [14, 45]). The amo operon thereby regulates nitrification and the enzyme ammonia monooxygenase (AMO), encoded by the amoABC cluster, catalyses the oxidation of ammonia to hydroxylamine that is further oxidized to nitrite, with help of the enzyme hydroxylamine oxidoreductase (HAO). Two c-type cytochromes, Cc554 and Ccm552, play a critical role in the electron transfer, whereby electrons are transported via Cc554 from HAO to Ccm552, to quinone or other terminal electron acceptors [17].

Major end product of the nitrification-denitrification cooperation is N_2, able to return to the atmosphere [78]. The enzymes, involved in the reduction of nitrate to N_2 or even to NH_3 work preferred under use of e^- and H^+, stored in organic matter, and in absence of oxygen, but a certain amount of oxygen is tolerated. Also in the dissimilatory denitrification process, a broad spectrum of bacteria, archaea and fungi is involved whereby their energy gaining strategy depends on various external factors such as climate, soil properties, and anthropogenic activities as TNF spreading [5]. So far, only partial sequences of the organisms, involved in the denitrification process are obtainable. Many of the bacteria beyond these genome sequences, which reduce nitrate in four steps over nitrite, nitric oxide and nitrous oxide to N_2, are not yet cultivable [4, 30].

The first reduction step from NO_3^- to NO_2^- is initiated by the nitrate reductase (Nar) enzyme complex, a two-electron reduction process. The nitrate reductases are molybdopterin enzymes and classified into three distinct types—respiratory (Nar), periplasmic (Nap), and assimilatory nitrate reductase (Nas)—defined by their cellular location, operon organization, and active site structure [78]. The molybdopterin oxide reductase Nar is located in the cytoplasm, whereas the nitrate reductase (Nap, NapAB) works in the periplasm [92]. At first nitrate must be transported into the cell that Nar can reduce NO_3^- to nitrite, that is afterwards transferred to the periplasm where it is reduced to NO and N_2O [46]. The nitric oxide reductases are key enzymes in the denitrification process encoded by nirK and nirS genes, which promote the reduction of nitrite to nitric oxide [9, 18, 126]. While the Nar genes regulate the nitrate respiration, nirK and nirS, encoding the copper-containing nitrite reductase genes, regulate the NO_2^- to NO reduction, whereby the transcription of nirK activates nirS. The nor genes encode the NO reductase and lead in an one-electron reduction step to N_2O. All these genes play periplasm located a crucial role in the denitrification process.

2.1 Nitrifying, Denitrifying Bacteria and Their Regulative

About 10–15% of the bacterial population, found in each known environmental niche, are able to nitrify and/or denitrify, but due to the difficulty in cultivating, most details are not yet known [4, 12, 22, 30, 126]. Nitrification occurs under aerobic, denitrification predominantly under anaerobic conditions. Thus, combining both processes in one single reactor is difficult to manage because of the different O_2 requirements. However, both energy gaining processes depend on each other and can thus coexist at reduced oxygen conditions. The regulation of denitrification starts by receiving transcription input signals from a sensor molecule. The absence of oxygen thereby equals an appropriate signal. Receiving such signal, a regulator protein binds to the promotor region and activates transcription, whereby the operator strength of transcription is influenced by the number of binding sites in the promotor region where contacts with the RNA polymerase are made [71].

Prerequisites for expression of denitrification genes are anaerobic conditions, the presence of NO, and the regulating nitrate reductase operon that is activated by the transcription regulator protein fumarate (Fnr), a one-component regulator of the nitrate reductase (Fig. 2; [71]). Sensor and regulator domain are within a single protein, the FNR/CRP protein, located in the promoter region of the denitrification

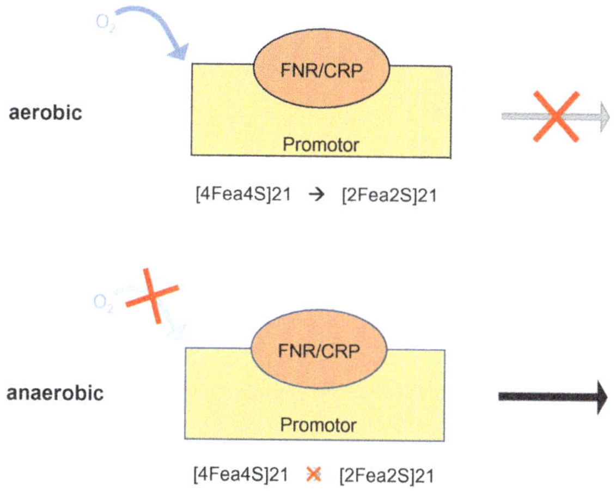

Fig. 2 The presence of oxygen destabilizes the [4Fe4S]21 cluster which causes the formation of the [2Fe2S]21 cluster. Under aerobic conditions, no FNR/CRP depending regulation takes place. Under anaerobic conditions, no formation to the [2Fe2S]21 cluster out of the [4Fe4S]21 cluster occurs, and FNR/CRP regulatory processes are conducted (after [104])

genes. During the shift from anaerobic to aerobic conditions, oxygen destabilizes the [4Fea4S]21 cluster of FNR/CRP and converts it to a [2Fea2S]21 cluster. Reducing conditions seem to be sufficient to reverse the process and FNR is converted to an active form to regulate anaerobic energy metabolism and induce the transcription of denitrification genes [71]. In *Pseudomonas stutzeri*, nitrate/nitrite is detected by a sensor transmitter in the periplasm that controls the expression of the *narX* gene. NarX is encoded by the narGHJI operon, a two-component regulatory system in the regulatory region of the *narL* encoded target operon [93]. However, other bacteria such as *Paracoccus denitrificans* do not have a *narXL* system but another nitrate regulatory mechanism, induced by Fnr-like regulatory proteins. In this case, DndR, Nnr and Anr proteins control the expression of the nitrate reductase during the denitrification process [104].

2.2 *N_2O Emission from Soils and WWTPs, a Problematic Side-Aspect of the Nitrification–Denitrification Process*

Denitrification may end incomplete, particularly when soil buffer capacity surpassing amounts of technically produced NH_4^+ are fertilized. In this scenario, denitrification involved microbes start saving energy and the reduction of NO_3^- to N_2 stops at the N_2O reduction level, since enough energy is already gained. The consequence is an increased emission of the intermediates NO and N_2O. This means that in N over fertilized soils, N_2O that is a highly efficient electron acceptor in gaining energy, is less reduced to N_2 in contrast to N shortage conditions to which soil inhabitants are

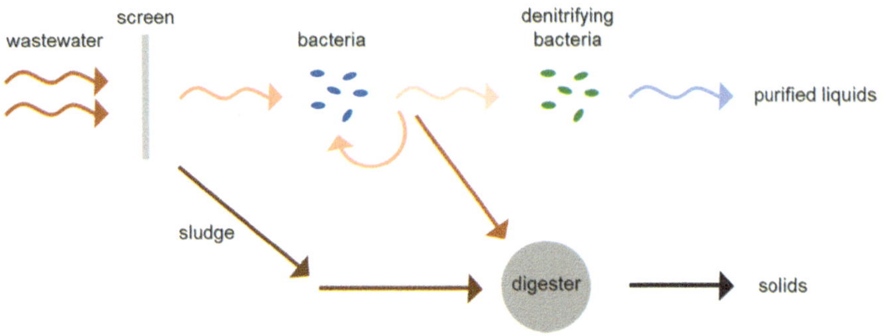

Fig. 3 Purification process of wastewater in sewage treatment plants. The first step is the removal of solids by a screen. Afterwards, the resulting sludge is transferred into a digester, while the liquid gets further purified. In the second step, bacteria and archaea remove organic substances. Remaining solids are recycled or discharged into the digester, which leads to final solids after further processing. During the third step, nitrate gets removed by denitrifying bacteria

long term evolutionary adapted ([13, 21]; Chapter "Soil Microbiological Recycling and the Virome Role in a Hectare Grassland"). Therefore, the consequence of N over-fertilization is enhanced emission of N_2O that is a 300-times more potent greenhouse gas than CO_2, strongly damaging the atmosphere in high levels [30, 67, 87, 97].

To recycle wastewater to reduce high organic loads, WWTPs are used. The first step in WWTPs is that wastewater passes a grit channel for removing large solid particles. Next, the wastewater enters an aeration tank in which the organic, largely N substituted load shall be oxidized to CO_2 and NO_3^- (Fig. 3). Subsequently, the wastewater passes an anaerobic zone in which the aerobically produced NO_3^- and toxic NO_2^- shall be removed with microbial help. The produced sewage sludge during the aerobic and anaerobic treatment is reduced in volume in a sewage sludge digester before being deposited or burnt. These purified wastewater have strict controls of carbon, nitrate, and phosphate loads to minimize eutrophication in the receiving waters. Intensively researched is furthermore how the wastewater can be further cleaned in tertiary wastewater purification units. Promising approaches are optimized autotrophic nitrifying and heterotrophic denitrifying bacteria and archaea processes at reduced aerobic conditions. It is further considered whether an efficient cooperation of both consortia under intermediate growth conditions would enable the removal of the produced NO_3^- under use of concomitantly enriching relatively recalcitrant organic wastewater loads as electron donor [12, 102]. Helpful in progressing are metagenome and 16S rRNA gene pyrosequencing for identifying the microflora including their process efficiency in soils and WWTPs. Gas chromatography (GC), high pressure liquid chromatography (HPLC) or fluorescent in situ hybridization (FISH) assays in combination with stable isotope spiking are performed to understand the biological degradation processes qualitatively and quantitatively [110, 123]. At present we know that the WWTPs core community consists mainly out of Proteobacteria. But also bacteria of the Actinobacteria, Bacteroidetes and Chloroflexi family, archaea, protozoa and fungi are involved in changing the oxygen availability, the wastewater salt content, the composition of the organic N substituted input, as well as qualitatively and quantitatively nitrate formation and reduction (Fig. 3; [36, 45, 113, 121]).

3 The potential of Synthetic Biology for efficient Nitrogen Recycling

3.1 Synthetic Biology Definition

Synthetic biology is an interdisciplinary field of science. Its aim is to design organisms with novel properties and metabolic pathways under use of genome-wide association studies and statistic Big Data analysis [1, 33, 37, 44, 108]. Microorganisms engineering has the advantage that useful bio-reagents can be more easily produced in a sustainable manner for human daily life. Additionally, it can be used to replace immense resources requiring techniques as technical N_2 fixing [23, 70] or to improve industrial and chemical processes in medicine, agriculture, and food processing [72]. Synthetic biology combines molecular biology and engineering along with predicting computational models. Thereby, modular DNA parts, regulatory DNA and RNA elements and coding sequences for proteins or terminators are used to create or optimize new biological systems [29, 60]. The acquisition of new properties does not just happen in the laboratory, but also in nature via horizontally transferred genes enabling organisms to adapt by outcompeting selection processes in distinct habitats. New denitrification properties enable a more efficient reduction of nitrate to N_2 by achieving a high energy output (see Sects. 4.3 and 6). The directed insertion of genes by means of synthetic biology may lead to an accelerating in achieving desired breeding successes with new properties. Within methods of molecular cloning as Golden Gate Assembly, it is possible to fuse several genes at one time and interconnect modular components for the creation of metabolic pathways as optimized denitrification [60, 68].

Particularly in cities with an increasing population and in areas with industrial livestock farming, huge amounts of wastewaters, sewage sludge, and animal manure are produced that generate high amounts of damaging greenhouse gases as N_2O and CH_4 [99, 105]. Thus, for advancing in knowledge concerning nitrifying and denitrifying bacteria and archaea, which in cooperation carry out a complete conversion of biologically and technically produced NH_3 into N_2, newly constructed microbes with surviving capabilities under competition by synthetic biology techniques can help solving inter alia the N_2O problem [16, 24, 73, 78, 81, 84, 119, 126]. Prominent representatives in solving the N_2O issue are N_2 producing bacteria as *Paracoccus stutzeri* or *Paracoccus denitrificans* and a decreased metabolism especially of autotrophic nitrifying and denitrifying bacteria and archaea at WWTP temperatures as low as 4 °C will pose a problem in growth efficacy and slow matter degradation [21, 51, 121]. It was observed by Yao et al. [121] in batch tests that some bacterial and archaeal strains are better adapted on varying temperatures in wastewaters with high organic C and N loads. For instance the bacteria *Janthinobacterium sp. M-11* and *Pseudomonas mohnii M-8* are cold-resistant heterotrophic-denitrification bacteria, capable to remove nitrogen at temperatures around 2 °C under aerobic and anaerobic conditions [120]. Furthermore, other temperature resistant nitrifying, denitrifying microbes, equipped with all enzymes needed for surviving under harsh soil and wastewater conditions by synthetic biology techniques offer a great potential to establish sustainable pathways for increasing the efficacy of N removal. Promising approaches in liberating soils and wastewaters from TNF overloads may be achievable by means of bio-engineering [31, 38, 77, 112, 118].

3.2 Application Possibilities of Synthetic Biology Techniques

In many parts of the world, blue jeans are in fashion, but just a few people know that the blue colouring is harmful for people and environment [49]. Inhaling dye fumes leads to lung damage among the workers in textile plants and wasterwaters strongly contribute to pollutions in nearby waters as an increase in biochemical and chemical oxygen demand or an impairment of photosynthesis [25]. Bacteria that produce high amounts of the dying colour or bacteria that can be used as detectors towards textile-colour-based pollution demonstrate promising approaches to counteract textile plant relates issues [28]. Ongoing attempts to modify microorganisms also deal with health (e.g. vaccine design), with the preserving of environments (e.g. biosensor development to detect and remove pollutants) or with a revolutionary bio-economy inter alia by energy saving via a more sustainable biofuel production for human welfare [32, 60]. The community behind genetic engineering is increasing and the biggest competition worldwide in synthetic biology is called iGEM (international Genetic Engineering Machine) [10]. Students from all over the world are challenged to create an idea to solve a "real-world problem" with means of synthetic biology and realise it within one year. All teams have to send in biological constructs, which have to be adapted to the BioBrick standard for an easy handling. The advantage of these standardized biological parts is the possibility to arrange them in an optional order to engineer new biological systems in the end. Therefore, the parts registry provided by iGEM works as a large modular system accessible for all prospective teams ([54]; Fig. 4).

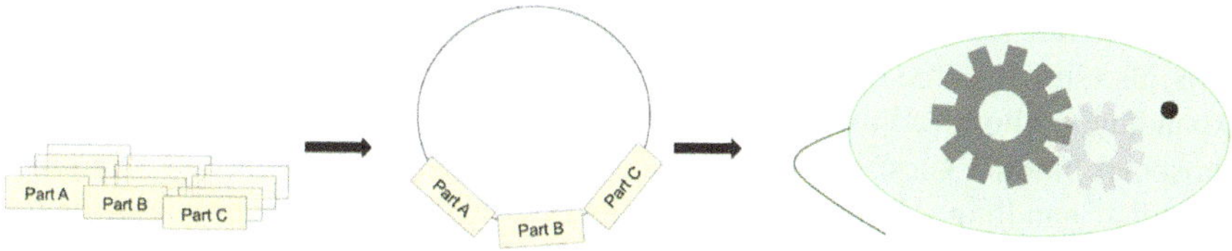

Fig. 4 Workflow of the generation of biological systems. Databases with standardized parts provide genetic elements with characteristic properties. Parts of interest can be chosen and fused into a vector by engineering a functional biological system. The created system gets implemented into a chassis to achieve new features

4 Engineered Bacteria for Optimizing Nitrogen Recycling

4.1 Common Cloning Methods in Synthetic Biology

Nitrogen-recycling concerned biology seeks strategies to find novel metabolic pathways for biomass production by concomitantly improving N input and N_2 output processes [16, 39, 84]. The traditional cloning works with restriction enzymes that cut specific sites in the DNA. In a second step, the insert is ligated within the plasmid. For accelerating cloning, Gibson Assembly and Golden Gate Assembly are promising tools, allowing the insertion of several genes just in one step. Gibson Assembly involves a T5 exonuclease, a Physios DNA polymerase, and a Tag DNA ligase. Overhangs in the DNA have to be generated that are chewed back from the exonuclease on the 5′ end, generating 3′ overhangs. The complementary DNA inserts can bind, and the polymerase subsequently fills the gaps. The ligase finally joins the complementary DNA fragments and seals the nicks [63, 64]. Regarding Golden Gate Assembly, this method is based on type II endonucleases that do not have the same recognition site and cutting site and distinguish them from conventional restriction enzymes. This feature allows the generation of non-palindromic base-overhangs, which can be ligated scar-less. There is a strict order of the cloned parts, determined by their base-overhangs. The final ligation is done by the T4 polymerase that is able to perform efficient end-joining [109].

4.2 Potentials of Chassis Organisms in Synthetic Biology

For the implementation of metabolic pathways recombinant model organisms, so called chassis, are required. An ideal chassis should meet many requirements necessary for the experimental setup, such as enough knowledge on the organism, an easy handling, and fast growth rates [20]. The gut bacterium *Escherichia coli (E. coli)* is a common example for chassis engineering. According to its well-known genetic background, low nutrient requirements, high growth rates and because of its cost effectiveness, *E. coli* is a popular organism for industrial and medical applications, also since desired results are reached with minimal effort [3, 20]. Furthermore, expanding toolboxes adjusted to *E. coli* that include genetic elements as promotors, ribosomal binding sites, or plasmidial origins of replication, ensure an enormous simplified handling. Human insulin production for instance is nowadays mainly conducted in *E. coli* [20]. One approach for the insulin production involves expression of chemically synthesized cDNA encoding for human proinsulin, including purification steps and excision of C-peptides afterwards. According to the increasing demand of insulin, optimization in chassis engineering is strongly investigated for improving the recombinant production [8].

For becoming an advantageous chassis, a selected organism of interest has to be precisely investigated and optimized. The genome has to be sequenced and genes have to be exactly annotated. One approach for chassis design is the top-down approach in which the removal of non-essential genes takes place. The principle is that chassis designed for specific production purposes contain only essential information for its function as growing or expressing crucial genes (Fig. 5). The advantage of such organisms is their functional metabolic network and an increased effectiveness in synthesizing the product of interest [24]. The design of such chassis can be assisted by computational analyses, predicting experimental outcomes, and by identifying the genes which are indispensable for the organism. With a minimal genome chassis, easier predictions can be made due to the decreased genome complexity [20]. The new DNA design can be reapplied to the host (e.g. *E. coli*) and eventual new generated features of the organism can be investigated [24]. The total synthesis or bottom up of the genome is another approach representing a milestone in synthetic biology. The genome of the hepatitis C virus (HCV) was the first completely synthesized genome in 2002 [26]. In 1995, the minimal genome project under the leadership of Craig Ventor was initiated with the aim to create an organism with the smallest amount of genes, the so called minimal genome, that is still able to survive [50]. A minimal genome has the potential to be the basis for the creation of further chassis. Thus, it can act as a universal chassis that enables the design of organisms with a higher complexity that match to specific experimental setups. *Mycoplasma genitalium*, the bacterium that Venter investigated, consists of only 525 genes and represents the smallest known bacterial genome up to now. A de novo synthesized, 1000.000 base pairs (bp) containing *M. mycoides* version was created by the transfer into the DNA-free bacterium *Mycoplasma capricolum*, leading to the novel strain JCVI-syn1.0. In 2016, the Craig Venter group was able to create the smallest, self-replicating *M. capricolum*, JCVI-syn3.0 with 473 genes, that could maintain cellular viability upon new genetic instructions by indicating the successful progress of the minimal genome project [26]. Because of natural limitations as host growth rates or biomass-production properties, it is very difficult to design an optimal host chassis. *Vibrio natriegens*, the fastest growing bacterium at lowest nutrients requirements is presently a promising candidate for an ideal chassis [20]. *V. natriegens* with a two-fold faster doubling time as *E. coli* enables an accelerated cloning and one attempt to create an optimized chassis with *V. natriegens* was undertaken by the iGEM team Marburg in

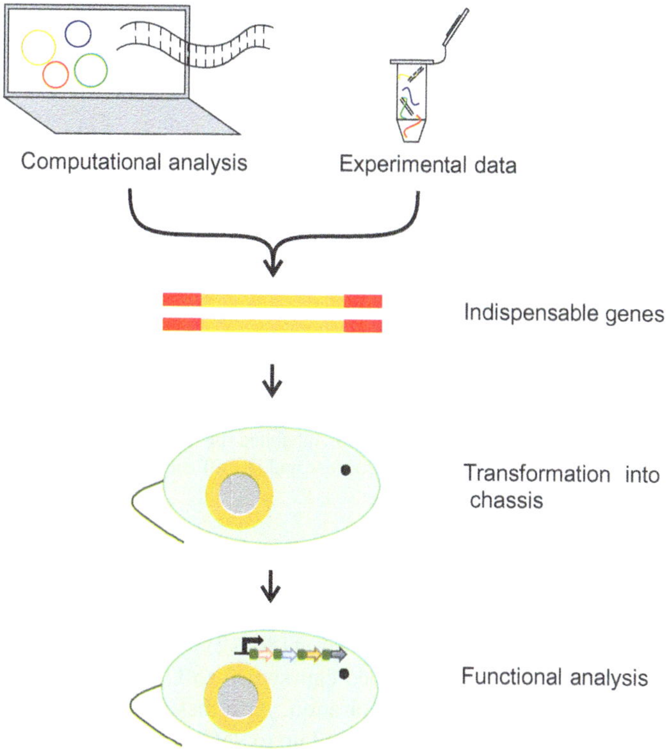

Fig. 5 The general principle of the top-down approach is to generate or optimize a chassis by using experimental data and computational analysis to detect indispensable genes. To achieve a reduced genome, analysed indispensable genes can be deleted. The newly designed system can be transferred into the host for approaching a chassis with altered properties that can be further studied

2018 [53], using natural competence regulators for introducing genetic modifications [106].

4.3 *Engineering Bacteria for Improving Nitrogen Recycling*

N-removal from soils and wastewaters requires environmental conditions for nitrifying and denitrifying bacteria and archaea. Thus, upon different O_2 requirements, denitrification and nitrification mostly occurs in separated areas as in soils and waters or different wastewater treatment reactors. Both processes can also occur in one system, called simultaneous nitrification and denitrification (SND), e.g. in suboptimal aerated tanks that do not exceed a distinct O_2 threshold. SND allows both processes to take part in one system and also occurs naturally in aeration tanks upon flocculation of microorganism. Dependent on floc size and concentration of dissolved

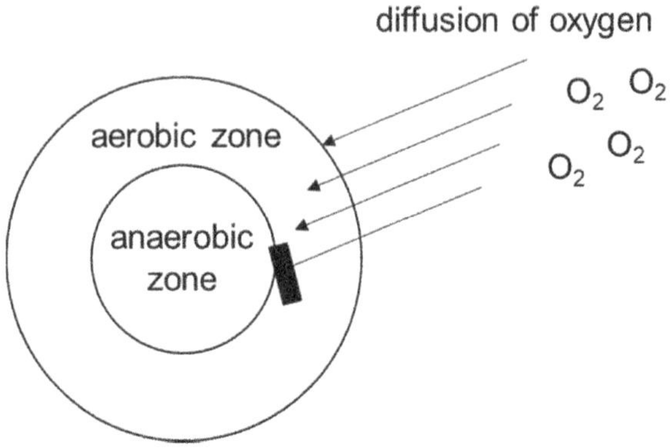

Fig. 6 A microbial sewage sludge floc with an outside aerobic zone, inhabited by O_2 respiring microorganisms and an anaerobic inside area, inhabited by anaerobic (NO_3^-) respiring and fermenting bacteria and archaea, which are floc size and O_2 diffusion dependent active

oxygen, anoxic and aerobic zones are everywhere present, in particularly in biofilms or sewage sludge flocs (Fig. 6; [35]). This phenomenon occurs due to limitations in oxygen diffusion, creating a microbial floc with an anoxic centre allowing denitrification and an aerobic coat allowing nitrification [86, 121]. At a high dissolved oxygen level, denitrifying bacteria and archaea can switch to O_2 respiration and thereby gain more energy than by nitrate respiration. At a dissolved O_2 concentration of around 0.5 mg/l, SND can be sustained and up to 95% of the available nitrogen can be recycled [65, 79, 82]. SND in wastewater treatment would have the advantage to simplify the overall process design by reducing total costs. Engineering a SND-performing organism by transferring genes responsible for nitrification to a denitrifying bacterium could be a solution to crown success in combining both capabilities to a single microbe that are useful in soil or wastewater treatment tanks. Both processes could be regulated by distinct promotors, which can be induced upon oxygen to ensure a temporal separation [11]. Under aerobic conditions, the genes for nitrification could be switched on and under anaerobic conditions the genes responsible for denitrification would be induced. The iGEM team Virginia from 2017 presented a project dealing with this idea [107]. *Paracoccus denitrificans* (*P. denitrificans*), that can survive under aerobic and anaerobic conditions, could be a well suitable candidate for a chassis that combines the genes for denitrification and nitrification. Therefore, the iGEM Team Virginia designed a construct with the crucial genes from the nitrifier *Nitrosomonas europaea* cloned into the genome of *P. denitrificans* (Fig. 7). The *P. denitrificans* construct carries the genes for nitrification and possesses the entire amo-operon encoding for the ammonia monooxygenase (AMO), HAO encoding for the hydroxylamine oxidoreductase, and the genes Cc554 and Ccm552 encoding for the electron transport responsible Cytochrome A and Cytochrome X. For the regulation of the system, the oxygen-dependent VHb gene (VHb) promoter was chosen of which the induction starts at dissolved oxygen concentrations of less than 2%. A tight separation of denitrification and nitrification ensures that nitrification stops

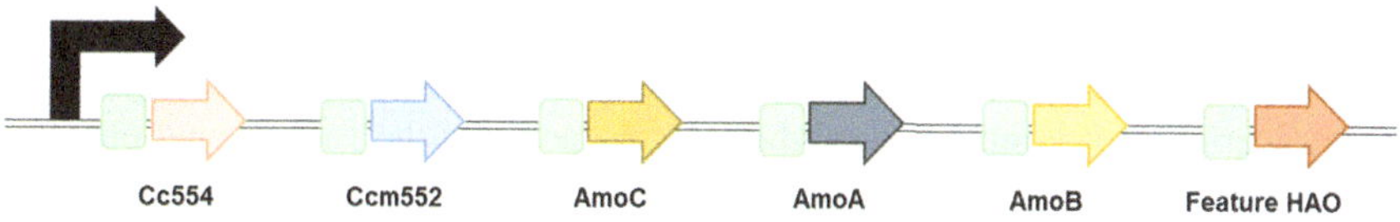

Fig. 7 A denitrifying genetic *P. denitrificans* construct, in which with synthetic biological methods genes from *N. europaea* are introduced. The system consists of a promotor at the beginning (black arrow), the ribosomal binding sites (green) the amo-operon (AmoA, AmoB, AmoC), and the genes Cc554, Ccm552 and HAO [52]

under anaerobic conditions [58]. Another promising candidate for future improving SND is *Thiosphaera pantotropha* (*T. pantotropha*). *T. pantotropha* is able to perform nitrification, denitrification and fix CO_2 with the special characteristic to be able to grow on a wide range of substrates as sulfur compounds or hydrogen. According to high growth rates, *T. pantotropha* is an attractive bacterium for efficient wastewater treatment [42]. Further characterization of the enzymes that are involved in the metabolic pathway and redesigning the properties by means of synthetic biology would make *T. pantotropha* the perfect chassis for SND (Fig. 7).

5 The Anammox Process

Certain planctomycetes were discovered in the early 1990s in wastewater sludge, which can oxidize ammonium with NO_2^- to dinitrogen gas [74]. This process is called anaerobic ammonium oxidation (anammox), which led to the term anammox bacteria for bacteria capable of performing this reaction. Some representatives of anammox bacteria are "*Candidatus* Kuenenia", "*Candidatus* Brocadia", "Anammoxoglobus", "*Candidatus* Jettenia" [47, 102]. During the anammox reaction, nitrite serves as electron acceptor and is reduced at first to nitric oxide. From the nitric oxide and ammonium as electron donor, the biosynthesis of hydrazine occurs, which is then further converted into dinitrogen gas via dehydration [56, 76]. The process thereby takes place in the so called anammoxosome, a membrane-bound intracytoplasmic cellular compartment [103]. Due to the ability to oxidize ammonium with nitrate or nitrite anaerobically, the anammox bacteria supplement the common nitrate cycle. These potentials also qualify the bacteria as promising candidates for improving wastewater treatment. In aerated wastewater treatment reactors or in aerated soils, fertilized with nitrate, anammox bacteria in combination with heterotrophic denitrifying bacteria and archaea can help reducing the biological oxygen demand (BOD) and enhancing organic carbon degradation [48]. Thereby, synthetic biology may further improve the procedure. For instance, *Nitrosomonas europaea*, a prevailing ammonia oxidizing bacterium, could be converted to a planctomycete lineage that grows well in synthetic medium with bicarbonate as only carbon source or in wastewater treatment reactors [88, 94, 101]. In general, anammox planctomycetes occur in a high number in freshwater samples and molecular biologically analysed wastewaters and can be auspicious chassis sources for designers, interested in reducing N overloads in WWTPs and soils. Anammox bacteria produce less sewage sludge, are thus costs lowering and promising candidates for reducing in the various environments NH_4^+, NO_3^-, and NO_2^- when skilled constructed with synthetic biological techniques [55, 94, 95, 101].

6 Future Perspectives

Synthetic biology is a logical extension of what is known as recombinant DNA (rDNA) technology or genetic engineering, that progresses continuously since the 1970s [57, 91]. Promising microorganisms for N concerned sustainable biotechnologies are also cyanobacteria, requiring a radical re-engineering with synthetic biology techniques to unlock their application potential [85]. Increasingly available devices for modifying cyanobacteria and their metabolic strategies include design advances concerning genetic promoters, ribosome binding sites, riboswitches, reporter proteins, modular vector systems, and markerless selection systems. Using new toolkits, cyanobacteria have been successfully engineered and express heterologous pathways for the production of a wide variety of valuable compounds [85]. For developing accurate predicting models for the best possible integration of pathways into the larger cellular metabolic network, in real-world applications afford the refinement of genetic circuits used for expressing heterologous pathways. However, it is obvious that despite the progress in synthetic biology and metabolic engineering in the last years, further improvements are required to make cyanobacteria fit and competitive for survival among heterotrophic microorganisms.

For a better N conversion performance, an improved elimination of surplus nitrate would be beneficial, when an engineered nitrifying and denitrifying chassis additionally possesses the capability to form a biofilm surrounded with extra polymeric substances (EPS) and structured with tubular pores, as the nitrifying bacterium *N. europaea* and the denitrifying bacterium *P. denitrificans* are demonstrating [35, 59]. Similar shaped as biofilms with tubular pore structures are sewage flocs, which ensure aerobic conditions in the outer region and anaerobic inside zones and work as a microbial fuel cell (MFC; Fig. 8). Sewage flocs help circumventing stress and synthetic biology tries combining the multicellular nitrifying and denitrifying potential of a sewage floc in one bacterium chassis. It is thereby of interest that the newly modelled chassis, integrated into a sewage floc, functions and exchanges its genetic support components and resources within the polymeric, compartmentalized, tubular sewage floc structure [100]. Such nitrification and denitrification combining cells could also be used as a MFC to generate electricity asides removing nitrate and nitrite. This idea has attracted much attention [66, 85, 122, 124]. A higher voltage in the cathode chamber is achieved by heterotrophic denitrifying MFC bacteria when dissolved oxygen in the surrounding high ammonia concentrated wastewater is high, between 3.0 and 4.2 mg/L, whereas total nitrogen (TN) removal is favoured at low dissolved oxygen levels (0.5–1.0 mg/L) [124]. Electrochemically active bacteria generate an

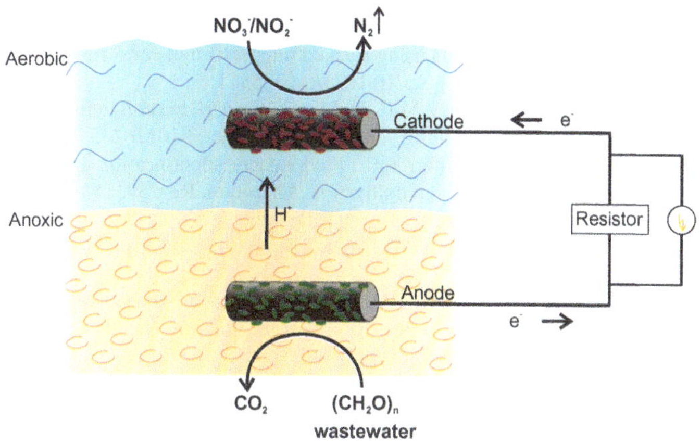

Fig. 8 The working of a bioelectrode, a microbial fuel cell (MFC): at the anode, CO_2 is reduced to biomass with water or NH_4^+ derived electrons and from the anode to the cathode electrons are pushed to reduce NO_2^- or NO_3^- by gaining energy, generating electricity. As anode and cathode are connected with a resistor to close the circuit in a MFC, nitrates and nitrites can act as e-acceptor and are reduced to molecular nitrogen [2]

electric current by pushing electrons from the anode towards the cathode. Nitrate and nitrite in flowing wastewater around the cathode or in soils act as electron acceptors [2, 122]. If the dissolved oxygen concentration is increased in a growth system with autotrophic, thermophilic nitrifying and facultative heterotrophic denitrifying bacteria population adapted to a low dissolved oxygen concentration, the bacterial composition apparently changes from a *Firmicutes* to a *Proteobacteria* dominance, from *Bacilli* to *alpha, beta*, and *gamma Proteobacteria* dominance [62]. In order to unravel such fundamental principles of life and to be able to improve applications in health, medicine, agriculture, veterinary as well as food industry, the engineering and continuous modification of microbial chassis, inter alia towards cathode, anode double-chamber with nitrifying and denitrifying MFC bacteria chassis, is presently one of the best ways to solve challenges of N overfertilization, wastewater purification and atmospheric pollution (Fig. 8; [24, 57, 75, 85, 125]).

References

1. Acevedo-Rocha, C.G.: The synthetic nature of biology. In: Hagen, K., Engelhard, M., Toepfer, G. (eds.) Ambivalences of Creating Life. Ethics of Science and Technology Assessment. Schriftenreihe der EA European Academy of Technology and Innovation Assessment GmbH, vol. 45, Springer, Cham (2016). https://doi.org/10.1007/978-3-319-21088-9_2

2. Aelterman, P., Rabaey, K., Pham, H.T., Boon, N., Verstraete, W.: Continuous electricity generation at high voltages and currents using stacked microbial fuel cells. Environ. Sci. Technol. **40**, 3388–3394 (2006). https://doi.org/10.1021/es0525511
3. Allocati, N., Masulli, M., Alexeyev, M.F., Di Ilio, C.: *Escherichia coli* in Europe: an overview. IJERPH **10**, 6235–6254 (2013). https://doi.org/10.3390/ijerph10126235
4. Amann, R.I., Ludwig, W., Schleifer, K.H.: Phylogenetic identification and in situ detection of individual microbial cells without cultivation. Microbiol. Rev. **59**, 143–169 (1995)
5. Ambus, P., Zechmeister-Boltenstern, S.: Denitrification and N-cycling in forest ecosystems. In: Bothe, H., Ferguson, S.J., Newton, W.E. (eds.) Biology of the Nitrogen Cycle, pp. 343–358. Elsevier (2007)
6. Arbuckle, T.E., Sherman, G.J., Corey, P.N., Walters, D., Lo, B.: Water nitrates and CNS birth defects: a population-based case-control study. Arch. Environ. Health **43**, 162–167 (1988). https://doi.org/10.1080/00039896.1988.9935846
7. Ashok, V., Hait, S.: Remediation of nitrate-contaminated water by solid-phase denitrification process—a review. Int. Environ. Sci. Pollut. Res. **22**, 8075–8093 (2015). https://doi.org/10.1007/s11356-015-4334-9
8. Baeshen, N.A., Baeshen, M.N., Sheikh, A., Bora, R.S., Ahmed, M.M.M., Ramadan, H.A.I., Kulvinder Singh Saini, K.S., Redwan, E.M.: Cell factories for insulin production. Microb. Cell Fact. **13**, S. 141 (2014). https://doi.org/10.1186/s12934-014-0141-0
9. Bakken, L.R., Bergaust, L., Liu, B., Frostegård, A.: Regulation of denitrification at the cellular level: a clue to the understanding of N_2O emissions from soils. Philos. Trans. R. Soc. Lond. B Biol. Sci. **367**, 1226–1234 (2012). https://doi.org/10.1098/rstb.2011.0321
10. Balmer, Andrew S., Bulpin, Kate, J.: Left to their own devices: Post-ELSI, ethical equipment and the international genetically engineered machine (iGEM) competition. BioSocieties **8**(3),311–335 (2013)
11. Baumann, B., Snozzi, M., Zehnder, A.J., van der Meer, J.R.: Dynamics of denitrification activity of *Paracoccus denitrificans* in continuous culture during aerobic-anaerobic changes. J. Bacteriol. **178**, 4367–4374 (1996). https://doi.org/10.1128/jb.178.15.4367-4374.1996
12. Benckiser, G.: Plastics, micro- and nanomaterials, and virus-soil microbe-plant interactions in the environment. In: Prasad, R. (ed.) Plant Nanobionics, Nanotechnology in the Life Sciences. Springer Nature, Switzerland (2019). https://doi.org/10.1007/978-3-030-12496-04
13. Benckiser, G., Haider, K., Sauerbeck, D.: Field measurements of gaseous nitrogen losses from an Alfisol planted with sugar beets. Z. Pflanzenernaehr. Bodenkd. **149**, 249–161 (1986)
14. Benckiser, G., Ladha, J.K., Wiesler, F.: Climate change and nitrogen turnover in soils and aquatic environments. In: Marxsen, J. (ed.) Climate Change and Microbial Ecology: Current Research and Future Trends, pp. 113–136. Caister Academic Press, UK (2016). https://doi.org/10.21775/9781910190319.08
15. Benckiser, G., Eilts, R., Linn, A., Lorch, H.-J., Sümer, E., Weiske, A., Wenzhofer, F.: N_2O emissions from different cropping systems and from aerated, nitrifying and denitrifying tanks of a municipal waste water treatment plant. Biol. Fertil. Soils. **23**(3), S. 257–265 (1996). https://doi.org/10.1007/BF00335953
16. Berla, B.M., Saha, R., Immethun, C.M., Maranas, C.D., Moon, T.S., Pakrasi, H.B.: Synthetic biology of cyanobacteria: unique challenges and opportunities. Front. Microbiol. (2013). https://doi.org/10.3389/fmicb.2013.00246
17. Bothe, H., Jost, G., Schloter, M., Ward, B.B., Witzel, K.: Molecular analysis of ammonia oxidation and denitrification in natural environments. FEMS Microbiol. Rev. **24**, 673–690 (2000). https://doi.org/10.1111/j.1574-6976.2000.tb00566.x
18. Braker, G., Zhou, J., Wu, L., Devol, A.H., Tiedje, J.M.: Nitrite reductase genes (nirK and nirS) as functional markers to investigate diversity of denitrifying bacteria in Pacific northwest marine sediment communities. Appl. Environ. Microbiol. **66**, 2096–2104 (2000). https://doi.org/10.1128/aem.66.5.2096-2104.2000
19. Brender, J.D., Weyer, P.J., Romitti, P.A., Mohanty, B.P., Shinde, M.U., Vuong, A.M., Sharkey, J.R., Dwivedi, D., Horel, S.A., Kantamneni, J., Huber, J.C., Jr., Zheng, Q., Werler, M.M., Kelley, K.E., Griesenbeck, J.S., Zhan, F.B., Langlois, P.H., Suarez, L., Canfield, M.A.: Prenatal

nitrate intake from drinking water and selected birth defects in offspring of participants in the national birth defects prevention study. Environ. Health Perspect. **121**, 1083–1089 (2013). https://doi.org/10.1289/ehp.1206249
20. Calero, P., Nikel, P.I.: Chasing bacterial chassis for metabolic engineering: a perspective review from classical to non-traditional microorganisms. Microb. Biotechnol. **12**, 98–124 (2019). https://doi.org/10.1111/1751-7915.13292
21. Carlson, C.A., Ingraham, J.L.: Comparison of denitrification by *Pseudomonas stutzeri*, *Pseudomonas aeruginosa*, and *Paracoccus denitrificans*. Appl. Environ. Microbiol. **45**, 1247–1253 (1983)
22. Casella, S., Payne, W.J.: Potential of denitrifiers for soil environment protection. FEMS Microbiol. Lett. **140**, 1–8 (1996). https://doi.org/10.1111/j.1574-6968.1996.tb08306.x
23. Chen, J.G., Crooks, R.M., Seefeldt, L.C., Bren, K.L., Bullock, R.M., Darensbourg, Y., Holland, P.L., Hoffman, B., Janik, M.J., Jones, A.K., Kanatzidis, M.G., King, P., Lancaster, M., Lymar, S.V., Pfromm, P., Schneider, W.F., Schrock, R.R.: Beyond fossil fuel–driven nitrogen transformations. Science **360**, 6391 (2018). https://doi.org/10.1126/science.aar6611
24. Chi, H., Xiaoli, W., Yue, S., Ying, Q., Zixin, D., Lianrong, W., Shi, C.: Engineering and modification of microbial chassis for systems and synthetic biology. Synth. Syst. Biotechnol. **4**, 25–33 (2019). https://doi.org/10.1016/j.synbio.2018.12.001
25. Cíntia, B.L., Fávaro-Polonio, Z., Pamphile, J.A., Polonio, J.C.: Effects of textile dyes on health and the environment and bioremediation potential of living organisms. Biotechnol. Res. Innov. **3**(2), 275–290 (2019). https://doi.org/10.1016/j.biori.2019.09.001
26. Coyle, M., Hu, J., Gartner, Z.: Mysteries in a minimal genome. ACS Sci. **2**, 274–277 (2016). https://doi.org/10.1021/acscentsci.6b00110
27. Croen, L.A., Todoroff, K., Shaw, G.M.: Maternal exposure to nitrate from drinking water and diet and risk for neural tube defects. Am. J. Epidemiol. **153**, 325–331 (2001). https://doi.org/10.1093/aje/153.4.325
28. Crow, D.: 6 Amazing things to watch in synthetic biology (2017). Online verfügbar unter https://medium.com/neodotlife/6-things-to-watch-in-synthetic-biology-f76666c7114e
29. Davies, J.A.: Real-world synthetic biology: is it founded on an engineering approach, and should it be? Life **9**, 6 (2019). https://doi.org/10.3390/life9010006
30. Demanèche, S., Philippot, L., David, M.M., Navarro, E., Vogel, T.M., Simonet, P.: Characterization of denitrification gene clusters of soil bacteria via a metagenomic approach. Appl. Environ. Microbiol. **75**, 534–537 (2009). https://doi.org/10.1128/AEM.01706-08
31. Di Capua, F., Pirozzi, F., Lens, P.N.L., Esposito, G.: Electron donors for autotrophic denitrification. Chem. Eng. J. **362**, 922–937 (2019). https://doi.org/10.1016/j.cej.2019.01.069
32. El Karoui, M., Hoyos-Flight, M., Fletcher, L.: Future trends in synthetic biology—a report. Front. Bioeng. Biotechnol. (2019). https://doi.org/10.3389/fbioe.2019.00175
33. Erb, T.J.: Neue biosynthetische Ansätze zur Kraft- und Wertstoffgewinnung. Nova Acta Leopold. NF Nr. **424**, 121–135 (2019)
34. FAO: World Fertilizer Trends and Outlook to 2020 (2017). www.fao.org
35. Flemming, H.C., Wingender, J., Szewzyk, U., Steinberg, P.D., Rice, S.C., Kjelleberg, S.: Biofilms: an emergent form of bacterial life. Nat. Rev. Microbiol. **14**(563–600), 75 (2016). https://doi.org/10.1038/nrmicro.2016.94
36. Franklin, R.B., Morrissey, E.M., Morina, J.C.: Changes in abundance and community structure of nitrate-reducing bacteria along a salinity gradient in tidal wetlands. Pedobiologia **60**, 21–26 (2017). https://doi.org/10.1016/j.pedobi.2016.12.002
37. Friedrich, B.: Biomasse als erneuerbarer Energieträger. Nova Acta Leopold. NF Nr. **424**, 105–120 (2019)
38. Girvan, H.M., Munro, A.W.: Applications of microbial cytochrome P450 enzymes in biotechnology and synthetic biology. Curr. Opin. Chem. Biol. **31**, 136–145 (2016). https://doi.org/10.1016/j.cbpa.2016.02.018
39. Goold, H.D., Wright, P., Hailstones, D.: Emerging opportunities for synthetic biology in agriculture. Genes **9**, 341 (2018). https://doi.org/10.3390/genes9070341

40. Greer, F.R., Shannon, M.: Infant methemoglobinemia: the role of dietary nitrate in food and water. Pediatrics **116**, 784–786 (2005). https://doi.org/10.1542/peds.2005-1497
41. Guidelines for Drinking-Water Quality, 4th edn.: World Health Organization, Geneva (2017)
42. Gupta, A. B.: Thiosphaera pantotropha: a sulphur bacterium capable of simultaneous heterotrophic nitrification and aerobic denitrification. Enzyme Micro. Technol. **21**(8), S. 589–595 (1997). https://doi.org/10.1016/S0141-0229(97)00070-7
43. Gupta, S.K., Gupta, C., Gupta, A.B., Seth, A.K., Bassin, J.K., Gupta, A.: Recurrent acute respiratory tract infections in areas with high nitrate concentrations in drinking water. Environ. Health Perspect. **108**, 363–366 (2000). https://doi.org/10.1289/ehp.00108363
44. Hahn, J., Katja Becker, K.: Dimensionen der antimikrobiellen Resistenz. Nova Acta Leopold. NF Nr. **424**, 147–165 (2019)
45. Halling-Sorensen, B., Jorgensen, S.E.: The removal of nitrogen from wastewater. In: Studies in Environmental Science, vol. 54, 1st edn., pp. 43–53. Elsevier (1993). https://doi.org/10.1016/S0166-1116(08)70524-7
46. Härtig, E., Schiek, U., Vollack, K.U., Zumft, W.G.: Nitrate and nitrite control of respiratory nitrate reduction in denitrifying *Pseudomonas stutzeri* by a two-component regulatory system homologous to NarXL of *Escherichia coli*. J. Bacteriol. **181**, 3658–3665 (1999)
47. Holmes, D.E., Dang, Y., Smith, J.A.: Nitrogen cycling during wastewater treatment. Adv. Appl. Microbiol. **106**, 113–192 (2019). https://doi.org/10.1016/bs.aambs.2018.10.003
48. Hossain, M.I., Paparini, A., Cord-Ruwisch, R.: Rapid adaptation of activated sludge bacteria into a glycogen accumulating biofilm enabling anaerobic BOD uptake. Bioresour. Technol. **228**, 1–8 (2017). https://doi.org/10.1016/j.biortech.2016.11.102
49. Hsu, T., Welner, D.H., Russ, Z.N., Cervantes, B., Prathuri, R.L., Adams, P.D., Dueber, J.E.: Employing a biochemical protecting group for a sustainable indigo dyeing strategy. Nat. Chem. Biol. **14**, 256–261 (2018). https://doi.org/10.1038/nchembio.2552
50. Hutchison, Clyde, A., Chuang, R,-Y., Noskov, Vladimir, N., Assad-Garcia, N., Deerinck, Thomas. J., Ellisman, M.H.. et al.: Design and synthesis of a minimal bacterial genome. In: *Science* (New York, N.Y.), vol. 351 (6280), aad6253. (2016). https://doi.org/10.1126/science.aad6253
51. Hwang, J.H., Oleszkiewicz, J.A.: Effect of cold-temperature shock on nitrification. Water Environ. Res. **79**, 964–968 (2007). https://doi.org/10.2175/106143007X176022
52. iGEM Team Virginia 2017; http://2017.igem.org/Team, Virginia (as of August 2020)
53. iGEM Team Marburg (2018) http://2018.igem.org/Team, Marburg (as of August 2020)
54. iGEM Foundation: iGEM Foundation. 27 Drydock Avenue, Suite 27E-230, Boston, MA 02210 (2019). https://2019.igem.org/Main_Page
55. Jetten, M.S.M., Sliekers, O., Kuypers, M., Dalsgaard, T., van Niftrik, L., Cirpus, I., van de Pas-Schoonen, K.T., Lavik, G., Thamdrup, B., Le Paslier, D., Op den Camp, H.J.M., Hulth, S., Nielsen, L.P., Abma, W., Third, K., Engström, P., Kuenen, J.G., Jørgensen, B.B., Canfield, D., Sinninghe-Damste, J.S., Revsbech, N.P., Fuerst, J., Weissenbach, J., Wagner, M., Schmidt, I., Schmid, M., Strous, M.: Anaerobic ammonium oxidation by marine and freshwater planctomycete-like bacteria. Appl. Microbiol. Biotechnol. **63**, 107–114 (2003). https://doi.org/10.1007/s00253-003-1422-4
56. Kartal, B., Almeida, N.M., Maalcke, W.J., Op den Camp, H.J.M., Jetten, M.S.M., Keltjens, J.T.: How to make a living from anaerobic ammonium oxidation. FEMS Microbiol. Rev. **37**(3), 428–461 (2013). https://doi.org/10.1111/1574-6976.12014
57. Katz, L., Chen, Y.Y., Gonzalez, R., Peterson, T.C., Zhao, H., Baltz, R.H.: Synthetic biology advances and applications in the biotechnology industry: a perspective. J. Ind. Microbiol. Biotechnol. **45**, 449 (2018). https://doi.org/10.1007/s10295-018-2056-y
58. Khosla, C., Bailey, J.E.: Characterization of the oxygen-dependent promoter of the *Vitreoscilla hemoglobin* gene in *Escherichia coli*. J. Bacteriol. **171**, 5995–6004 (1989). https://doi.org/10.1128/jb.171.11.5995-6004.1989
59. Kim, Y., Kang, J., Shen, B., Wang, Y., He, Y., Lee, M.: Open–closed switching of synthetic tubular pores. Nat. Commun. **6**, 8650 (2015). https://doi.org/10.1038/ncomms9650

60. König, H., Frank, D., Heil, R., Coenen, C.: Synthetic genomics and synthetic biology applications between hopes and concerns. Curr. Genomics **14**, 11–24 (2013). https://doi.org/10.2174/1389202911314010003
61. Landesumweltamt Brandenburg (LUA): Gefährdungsabschätzung und Sanierung von ehemaligen Rieselfeldern unter Berücksichtigung der Anforderungen von BBodSchG/BBodSchV; Fachbeiträge des Landesumweltamtes - Titelreihe, Heft - Nr. 77 - Bodenschutz und Altlastenbearbeitung 1, April 2003
62. Lee, H.W., Park, Y.K., Choi, E., Lee, J.W.: Bacterial community and biological nitrate removal: comparisons of autotrophic and heterotrophic reactors for denitrification with raw sewage sludge. J. Microbiol. Biotechnol. **11**, 1826–1835 (2008). https://doi.org/10.4014/jmb.0800.276
63. Li, L., Jiang, W., Lu, Y.: A modified Gibson assembly method for cloning large DNA fragments with high GC contents. Methods Mol. Biol. **1671**, 203–209 (2018). https://doi.org/10.1007/978-1-4939-7295-1_13
64. Li, S., Tian, Y., Wu, K., Ye, Y., Yu, J., Zhang, J., Liu, Q., Hu, M., Li, H., Tong, Y., Harberd, N., Fu, X.: Modulating plant growth-metabolism coordination for sustainable agriculture. Nature **560**, 595–600 (2018). https://doi.org/10.1038/s41586-0415-5
65. Lim, J.Y., Kim, H.S., Park, S.Y., Kim, J.H.: Simultaneous nitrification and denitrification by using ejector type microbubble generator in a single reactor. Environ. Eng. Res. **25**, 252–257 (2020). https://doi.org/10.4491/eer.2018.427
66. Lovely, D.R., Walker, D.J.F.: Geobacter protein nanowires. Front. Microbiol. (2019). https://doi.org/10.3389/fmicb.2019.02078
67. Lycus, P., Soriano-Laguna, M.J., Kjos, M., Richardson, D., Gates, A.J., Milligan, D.A., Frostegård, Å., Bergaust, L., Bakken, L.R.: A bet-hedging strategy for denitrifying bacteria curtails their release of N_2O. PNAS **115**, 11820–11825 (2018). https://doi.org/10.1073/pnas.1805000115
68. McCarty, N.S., Ledesma-Amaro, R.: Synthetic biology tools to engineer microbial communities for biotechnology. Trends Biotechnol. **37**, 181–197 (2019). https://doi.org/10.1016/j.tibtech.2018.11.002
69. Miao, L., Liu, Z.: Microbiome analysis and -omics studies of microbial denitrification processes in wastewater treatment: recent advances. Sci. China Life Sci. **61**(7), 753–761 (2018). https://doi.org/10.1007/s11427-017-9228-2
70. Michalsky, R., Pfromm, P.H., Steinfeld, A.: Rational design of metal nitride redox materials for solar-driven ammonia synthesis. Interface Focus **5**, 20140084 (2015). https://doi.org/10.1098/rsfs.2014.0084
71. Molecular Medical Microbiology: In: Tang, W., Sussman, M., Liu, D., Poxton, I., Schwartzman, J. (eds.). Elsevier ScienceDirect (2015). https://doi.org/10.1016/C2010-1-67744-9
72. Mortimer, J.C.: Plant synthetic biology could drive a revolution in biofuels and medicine. Exp. Biol. Med. **244**, 323–331 (2019). https://doi.org/10.1177/1535370218793890
73. Muck, S., De Corte, D., Clifford, E.L., Bayer, B., Gerhard, J., Herndl, G.J., Sintes, E.: Niche differentiation of aerobic and anaerobic ammonia oxidizers in a high latitude deep oxygen minimum zone. Front. Microbiol. (2019). https://doi.org/10.3389/fmicb.2019.02141
74. Mulder, A., Graaf, A.A., Robertson, L.A., Kuenen, J.G.: Anaerobic ammonium oxidation discovered in a denitrifying fluidized bed reactor. FEMS Microbiol. Ecol. **16**(3), 177–184 (1995). https://doi.org/10.1111/j.1574-6941.1995.tb00281.x
75. Nguyen, P.Q.: Synthetic biology engineering of biofilms as nanomaterials factories. Biochem. Soc. Trans. **45**, 585–597 (2017). https://doi.org/10.1042/BST20160348
76. Oshiki, M., Satoh, H., Okabe, S.: Ecology and physiology of anaerobic ammonium oxidizing bacteria. Environ. Microbiol. **18**(9), 2784–2796 (2016). https://doi.org/10.1111/1462-2920.13134
77. Peter, D.M., von Borzyskowski, L.S., Kiefer, P., Christen, P., Vorholt, J.A., Erb, T.J.: Screening and engineering the synthetic potential of carboxylating reductases from central metabolism and polyketide biosynthesis. Angew. Chem. **54**, 13457–13461 (2015)

78. Philippot, L.: Denitrifying genes in bacterial and Archaeal genomes. Biochim. Biophys. Acta (BBA) **1577**, 355–376 (2002). https://doi.org/10.1016/S0167-4781(02)00420-7
79. Pochana, K.: Study of factors affecting simultaneous nitrification and denitrification (SND). Water Sci. Technol. **39**, 61–68 (1999). https://doi.org/10.1016/S0273-1223(99)00123-7
80. Pomeranz, A., Korzets, Z., Vanunu, D., Krystal, H., Wolach, B.: Elevated salt and nitrate levels in drinking water cause an increase of blood pressure in schoolchildren. Kidney Blood Press. Res. **23**, 400–403 (2000). https://doi.org/10.1159/000025989
81. Prosser, J.I.: Nitrogen in soils. In: Nitrification Encyclopedia of Soils in the Environment, pp. 31–39 (2005). https://doi.org/10.1016/B0-12-348530-4/00512-9
82. Qureshi, N., Annous, B.A., Ezeji, T.C., Karcher, P., Maddox, I.S.: Biofilm reactors for industrial bioconversion processes: employing potential of enhanced reaction rates. Microb. Cell Fact. **4**, 24 (2005). https://doi.org/10.1186/1475-2859-4-24
83. Rádiková, Z., Tajtakova, M., KoCan, A., Trnovec, T., Sebokova, E., Klimes, I., Langer, P.: Possible effects of environmental nitrates and toxic organochlorines on human thyroid in highly polluted areas in Slovakia. Thyroid **18**, 353–362 (2008). https://doi.org/10.1089/thy.2007.0182
84. Rogers, C., Oldroyd, G.E.D.: Synthetic biology approaches to engineering the nitrogen symbiosis in cereals. J Exp. Bot. **65**, 1939–1946 (2014). https://doi.org/10.1093/jxb/eru098
85. Santos-Merino, M., Singh, A.K., Ducat, D.C.: New applications of synthetic biology tools for cyanobacterial metabolic engineering. Front. Bioeng. Biotechnol. (2019). https://doi.org/10.3389/fbioe.2019.00033
86. Satoh, H., Nakamura, Y., Ono, H., Okabe, S.: Effect of oxygen concentration on nitrification and denitrification in single activated sludge flocs. Biotechnol. Bioeng. **83**, 604–607 (2003). https://doi.org/10.1002/bit.10717
87. Schlesinger, W.H.: On the fate of anthropogenic nitrogen. PNAS **106**, 203–208 (2009). https://doi.org/10.1073/pnas.0810193105
88. Schmidt, I., Sliekers, O., Schmid, M., Cirpus, I., Strous, M., Bock, E., et al.: Aerobic and anaerobic ammonia oxidizing bacteria-competitors or natural partners? FEMS Microbiol. Ecol. **39**, 175–181 (2002). https://doi.org/10.1111/j.1574-6941.2002.tb00920.x
89. Schullehner, J., Hansen, B., Thygesen, M., Pedersen, C.B., Sigsgaard, T.: Nitrate in drinking water and colorectal cancer risk: a nationwide population-based cohort study. Int. J. Cancer **143**, 73–79 (2018). https://doi.org/10.1002/ijc.31306
90. Schwientek, M., Einsiedl, F., Stichler, W., Stögbauer, A., Strauss, H., Maloszewski, P.: Evidence for denitrification regulated by pyrite oxidation in a heterogeneous porous groundwater system. Chem. Geol. **255**, 60–67 (2008). https://doi.org/10.1016/j.chemgeo.2008.06.005
91. Shao, Y., Lu, N., Wu, Z., Cai, C., Wan, S., Zhang, L.L., Zhou, F., Xiao, S., Liu, L., Zeng, X., Zheng, H., Yang, C., Zhao, Z., Zhao, G., Zhou, J.Q., Xue, X., Qin, Z.: Creating a functional single-chromosome yeast. Nature **560**, 331–335 (2018). https://doi.org/10.1038/s41586-018-0382x
92. Sparacino-Watkins, C., Stolz, J.F., Basu, P.: Nitrate and periplasmic nitrate reductases. Chem. Soc. Rev. **43**, 676–706 (2014). https://doi.org/10.1039/c3cs60249d
93. Stewart, V., Parales, J.: Identification and expression of genes narL and narX of the nar (nitrate reductase) locus in *Escherichia coli* K-12. J. Bacteriol. **170**, 1589–1597 (1988). https://doi.org/10.1128/jb.170.4.1589-1597.1988
94. Strous, M., van Gerven, E., Zheng, P., Kuenen, J.G., Jetten, M.S.M.: Ammonium removal from concentrated waste streams with the anaerobic ammonium oxidation (Anammox) process in different reactor configurations. Water Res. **31**, 1955–1962 (1997). https://doi.org/10.1016/S0043-1354(97)00055-9
95. Suto, R., Ishimoto, C., Chikyu, M., Aihara, Y., Matsumoto, T., Uenishi, H., et al.: Anammox biofilm in activated sludge swine wastewater treatment plants. Chemosphere **167**, 300–307 (2017). https://doi.org/10.1016/j.chemosphere.2016.09.121
96. Tajtáková, M., Semanová, Z., Tomková, Z., Szökeová, E., Majoroš, J., Rádiková, Ž., et al.: Increased thyroid volume and frequency of thyroid disorders signs in school children from

nitrate polluted area. Chemosphere **62**, 559–564 (2006). https://doi.org/10.10.1016/j.-chemosphere.2005.06.030
97. Torres, M.J., Simon, J., Rowley, G., Bedmar, E.J., Richardson, D.J., Gates, A.J., Delgado, M.J.: Nitrous oxide metabolism in nitrate-reducing bacteria: physiology and regulatory mechanisms. Adv. Microb. Physiol. **68**, 353–432 (2016). https://doi.org/10.1016/bs.ampbs.2016.02.007
98. Tricker, A.R., Preussmann, R.: Carcinogenic N-nitrosamines in the diet: occurrence, formation, mechanisms and carcinogenic potential. Mutat. Res. Genet. Toxicol. **259**, 277–289 (1991). https://doi.org/10.1016/0165-1218(91)90123-4
99. Tumendelger, A., Alshboul, Z., Lorke, A.: Methane and nitrous oxide emission from different treatment units of municipal wastewater treatment plants in Southwest Germany. PLoS One **14**, e0209763 (2019). https://doi.org/10.1371/journal.pone.0209763
100. Uemoto, H., Saiki, H.: Nitrogen removal by tubular gel containing *Nitrosomonas europaea* and *Paracoccus denitrificans*. Appl. Environ. Microbiol. **62**, 4224–4228 (1996)
101. Van de Graaf, A.A., Bruijn, P., Robertson, L.A., Jetten, M.S.M., Kuenen, J.G.: Autotrophic growth of anaerobic ammonium-oxidizing micro-organisms in a fluidized bed reactor. Microbiology **142**, 2187–2196 (1996). https://doi.org/10.1099/13500872-142-8-2187
102. Van Niftrik, L., Jetten, M.S.M.: Anaerobic ammonium-oxidizing bacteria: unique microorganisms with exceptional properties. Microbiol. Mol. Biol. Rev. **76**, 585–596 (2012). https://doi.org/10.1128/MMBR.05025-11
103. van Niftrik, L., van Helden, M., Kirchen, S., van Donselaar, E.G., Harhangi, H.R., Webb, R.I., et al.: Intracellular localization of membrane-bound ATPases in the compartmentalized anammox bacterium '*Candidatus* Kuenenia stuttgartiensis'. Mol. Microbiol. **77**(3), 701–715 (2010). https://doi.org/10.1111/j.1365-2958.2010.07242.x
104. Van Spanning, R.J., Boer, A.P., Reijnders, W.N., Westerhoff, H.V., Stouthamer, A.H., van der Oost, J.: FnrP and NNR of *Paracoccus denitrificans* are both members of the FNR family of transcriptional activators but have distinct roles in respiratory adaptation in response to oxygen limitation. Mol. Microbiol. **23**, 893–907 (1997). https://doi.org/10.1046/j.1365-2958.1997.2801638.x
105. Velthof, G.L., Lesschen, J.P., Webb, J., Pietrzak, S., Miatkowski, Z., Pinto, M., et al.: The impact of the nitrates directive on nitrogen emissions from agriculture in the EU-27 during 2000–2008. Sci. Total Environ. **468–469**, 1225–1233 (2014). https://doi.org/10.1016/j.scitotenv.2013.04.058
106. VIBRIGENS—Accelerating SynBio. In: iGEM Team Marburg 2018 (2018). https://2018.igem.org/Team:Marburg/Team
107. Virginia iGEM Team 2017: Sewage PD—a single chassis ammonia removal device for use in wastewater treatment systems (2017). https://2017.igem.org/Team:Virginia
108. Visscher, P.M., Wray, N.R., Zhang, Q., Sklar, P., McCarthy, M.I., Brown, M.A., Yang, J.: 10 years of GWAS discovery: biology, function, and translation. Am. J. Hum. Genet. **101**, 5–22 (2017). https://doi.org/10.1016/j.ajhg.2017.06.005
109. Vladimir, P., Ong, J.L., Kucera, R.B., Langhorst, B.W., Bilotti, K., Pryor, J.M., et al.: Optimization of Golden Gate assembly through application of ligation sequence-dependent fidelity and bias profiling (1321) (2018)
110. Wagner, M.: Single-cell ecophysiology of microbes as revealed by Raman microspectroscopy or secondary ion mass spectrometry imaging. Annu. Rev. Microbiol. **63**, 411–429 (2009). https://doi.org/10.1146/annurev.micro.091208.073233
111. Wakida, F.T., Lerner, D.N.: Non-agricultural sources of groundwater nitrate: a review and case study. Water Res. **39**(1), 3–16 (2005). https://doi.org/10.1016/j.watres.2004.07.026
112. Wang, H.H.: Synthetic genomes for synthetic biology. J. Mol. Cell Biol. **2**, 178–179 (2010). https://doi.org/10.1093/jmcb/mjq015
113. Wang, Z., Zhang, X.X., Lu, X., Liu, B., Li, Y., Long, C., Li, A.: Abundance and diversity of bacterial nitrifiers and denitrifiers and their functional genes in tannery wastewater treatment plants revealed by high-throughput sequencing. PLoS One **9**(11), e113603 (2014). https://doi.org/10.1371/journal.pone.0113603

114. Ward, M.H., deKok, T.M., Levallois, P., Brender, J., Gulis, N.B.T., VanDerslice, J.: Workgroup report: drinking-water nitrate and health—recent findings and research needs. Environ. Health Perspect. **113**, 1607–1614 (2005). https://doi.org/10.1289/ehp.8043
115. Ward, M., Jones, R., Brender, J., Kok, T., Weyer, P., Nolan, B., Villanueva, C.M., van Breda, S.G.: Drinking water nitrate and human health: an updated review. IJERPH **15**, 1557 (2018). https://doi.org/10.3390/ijerph15071557
116. Warneke, S., Schipper, L.A., Bruesewitz, D.A., Baisden, W.T.: A comparison of different approaches for measuring denitrification rates in a nitrate removing bioreactor. Water Res. **45**, 4141–4151 (2011). https://doi.org/10.1016/j.watres.2011.05.027
117. Warneke, S., Schipper, L.A., Matiasek, M.G., Scow, K.M., Cameron, S., Bruesewitz, D.A., McDonald, I.R.: Nitrate removal, communities of denitrifiers and adverse effects in different carbon substrates for use in denitrification beds. Water Res. **45**, 5463–5475 (2011). https://doi.org/10.1016/j.watres.2011.08.007
118. Wentzel, M.C., Ekama, G.A., Marais, G.V.R.: Processes and modelling of nitrification denitrification biological excess phosphorus removal systems—a review. Water Sci. Technol. **25**, 59–82 (1992). https://doi.org/10.2166/wst.1992.0114
119. Wu, Y., Lu, L., Wang, B., Lin, N., Zhu, J., Cai, Z., Yan, X., Jia, Z.: Long-term field fertilization significantly alters community structure of ammonia-oxidizing bacteria rather than archaea in a paddy soil. SSSAJ **75**, 1431–1439 (2011). https://doi.org/10.2136/sssaj2010.0434
120. Yang, J., Kim, B., Kim, G.., Jung, G.Y., Seo, S.W.: Synthetic biology for evolutionary engineering: from perturbation of genotype to acquisition of desired phenotype. Biotech. Biofuels **12**, S. 113 (2019). https://doi.org/10.1186/s13068-019-1460-5
121. Yao, S., Ni, J., Ma, T., Li, C.: Heterotrophic nitrification and aerobic denitrification at low temperature by a newly isolated bacterium, *Acinetobacter* sp. HA2. Bioresour. Technol. **139**, 80–86 (2013). https://doi.org/10.1016/j.biortech.2013.03.189
122. Zhang, Y., Angelidaki, I.: Bioelectrode-based approach for enhancing nitrate and nitrite removal and electricity generation from eutrophic lakes. Water Res. **46**, 6445–6453 (2012). https://doi.org/10.1016/j.watres.2012.09.022
123. Zhang, L., Shen, Z., Fang, W., Gao, G.: Composition of bacterial communities in municipal wastewater treatment plant. Sci. Total Environ. **689**, 1181–1191 (2019). https://doi.org/10.1016/j.scitotenv.2019.06.432
124. Zhao, J., Wu, J., Li, X., Wang, S., Hu, B., Ding, X.: The denitrification characteristics and microbial community in the cathode of an MFC with aerobic denitrification at high temperatures. Front. Microbiol. **8**, 9 (2017). https://doi.org/10.3389/fmicb.2017.00009
125. Zhao, H., Zeng, An-Ping (Hg.): Synthetic Biology - Metabolic Engineering. 1st edition 2018. Cham: Springer International Publishing (Advances in Biochemical Engineering/Biotechnology) (2018).
126. Zumft, W.G.: Cell biology and molecular basis of denitrification. Microbiol. Mol. Biol. Rev. **61**, 533–616 (1997)

Plant Demand Adapted Fertilization in Organic and Precision Farming

D. L. N. Rao, P. Dey, and K. Sammi Reddy

Abstract The major agricultural challenges in the developing world are in ensuring food and nutritional security, improving soil health and maintaining environmental quality. Reduced usage of organics, imbalanced fertilizer use, inherent soil constraints and emerging nutrient deficiencies are the major contributors of poor soil health. The hotspots of N fertilizer application have shifted from the USA and Western Europe in the 1960s to eastern Asia in the early 21st century thus bringing soil sustainability concerns. Simultaneously, the interest in low external inputs or organic agriculture has increased. This review presents some the important lessons learnt in precision chemical fertilization and organic fertilization effects on productivity and soil health. Both approaches provide solutions to ensure improved fertilizer use efficiency in the short-term and maintain soil fertility in the long term. But organic farming generally lacks the potential for a short-term on-the-go reaction especially with respect to fertilizer application. It is argued that organo-mineral precision fertilization has the advantages of nutrient supply in sufficient quantities, nutrient efficiency and soil health improvement. Organo-mineral precision fertilization (OM-PF) requires long term research on nutrient availability, crop yields, effects on soil biology and fertility.

Keywords Fertilizers · Nitrogen · Nutrient Management · Plant Nutrition · Soil Health · Soil Testing · Sustainability

---

D. L. N. Rao (✉) · P. Dey
ICAR-Indian Institute of Soil Science, Bhopal, M.P., India
e-mail: dlnrao.icar@gmail.com

P. Dey
e-mail: pradipdey@yahoo.com

K. S. Reddy
ICAR-Central Research Institute for Dryland Agriculture, Hyderabad, India
e-mail: ksreddy_iiss39@yahoo.com

G. Benckiser (ed.), *Soil and Recycling Management in the Anthropocene Era*, Environmental Science and Engineering,
https://doi.org/10.1007/978-3-030-51886-8_6

1 Introduction

The major agricultural challenges in the developing world are ensuring food and nutritional security, improving soil health and maintaining environmental quality. Reduced organic waste recycling, low and imbalanced fertilizer use, not correcting inherent soil constraints or emerging nutrient deficiencies are the major contributors of poor soil health. Continued high fertilizer usage in post-green revolution era has also brought sustainability concerns. The hotspots of N fertilizer application have shifted from the USA and Western Europe in the 1960s to eastern Asia in the early twenty-first century. Europe, China, and India now account for over 50% of the N used [26, 54]. Together with the enhanced cropping of nitrogen fixing legumes, it has led to an expansion of the reactive N pool in the environment [28]. Inefficiencies in N uptake lead to significant N losses and associated environmental problems which are likely to be compounded by global climate change. Over 3/4ths of nitrous oxide emissions from Indian agricultural soils are from applied nitrogen fertilizers [67]. Low fertilizer nutrients use efficiency by crops, typically for N (30–50%), P (15–20%), S (8–12%), Zn (2–5%), Fe (1–2%) and Cu (1–2%) [61] has been attributed to deterioration in soil health and poor synchronization between crop demand for nutrients and their availability from soil.

Traditional farming is based on local practices and improving the fertilizer use efficiency will require transition to precision farming, which combines sensors, robots, drones, GPS, mapping tools and data-analytics software to customize the care that plants receive [49]. Data on individual plants like stem size, leaf shape and the moisture of the soil around a plant is fed to computers, which analyses for symptoms of stress and gives feedback to farmers in real time to deliver water, pesticide or fertilizer in required doses only. In map-based VRA technology a map is generated of the farm and fed into the computer on board that decides which fertilizer should be applied in a given parcel of land and in what quantity. Precision farming efficiently manages spatial variations in the field, reduces water and chemical use, and produces healthier crops and higher yields-all of which improves farmers' incomes, conserves resources and reduces chemical runoff. However, there are several challenges that limit its broader adoption: socio-economical, agronomical and technological. Precision farming technology has moderate risk-reduction benefits which lead to higher profits in the long run due to decreased variability in yields in a field. Precision agriculture (PA) is now being increasingly adopted in developed countries [37]; according to Molin and Tavares [58] about 15% of the grain producers in Brazil, use PA techniques to apply inputs at variable rates based on soil characteristics. In many developing countries with small size land holdings, other site-specific nutrient management techniques and drip irrigation/fertigation may be more simpler and economic alternatives that approach precision agriculture in efficiency.

There has been a worldwide upsurge on interest in organic farming. As of 2015, India had the most organic producers followed by Uganda and Mexico. In 2018, India was 9th in terms of land under organic agriculture [102] with 3.56 million ha under organic certification of which half was cultivated and half was wild harvest collection. Conventional agriculture relies on readily soluble forms of nutrients for nutrient supply while organic systems use a long-term strategy relying upon crop rotations, nitrogen fixing cover crops, and manures, compost and biofertilizers to derive nutrients through biological cycles to optimize crop production. The present review aims to present the most important lessons learnt in precision chemical fertilization and organic fertilization with a focus on developing countries and also discusses environmental influences including effects on soil health in the context of the current concerns.

2 Organic Farming Approaches

In organic systems, farmers manage the whole farm like a living organism with a balanced growth of crops and livestock holding to ensure that nutrient cycles are closed as far as possible. Only the nutrients in the form of food are exported out of the farm. Plant nutrition depends on "biologically derived" nutrients from organic materials; the challenge is to economically meet crop nutrient needs and address soil nutrient depletion. The benefits include lesser leaching losses of nitrogen due to extended mineralization period of organic matter and soil structure improvement and replenishment of the nutrient-holding capacity of the soil by building up microbial biomass.

2.1 Strategies to Meet Nutrient Demand

Manures contain nitrogen both in organic and ammonium forms; N mineralization from manures with a C:N ratio < 20:1 occurs readily; but only 15% of the organic N in the compost may be available during that season if ratio is high. Application of compost with C:N ratio > 30:1 will result in net immobilization of soil nitrogen with reduced nitrogen availability to the first crop. Some estimates of organic N availability for some of the manures through years 2 through year 4 are 15, 7 and 4% per year ([104]; Table 1). Succulent plant residues are mineralized faster than more lignified ones, so the type and timing of incorporation must be tailored to needs of the crop in question to create synchrony with crop demand. Incorporation of nutrient poor wide C:N ratio plant material such as cereal straw can serve to prevent leaching by immobilizing the nitrate N, thus ensuring also a build-up of microbial biomass which will be re-mineralized later to meet crop demand.

Cover crops improve soil physical properties and soil biological activity. They improve nutrient cycling by scavenging residual nitrogen mineralized from soil and organic amendments (before it is lost to volatilization, runoff, or leaching) and

Table 1 Nutrient analysis and approximate release speed (time required for mineralization of guaranteed nutrients) of common products for organic production

	% N	% P_2O_5	% K_2O	Release speed
Nitrogen				
Alfalfa meal	2	1	2	2–6 months
Blood meal	12	1	1	6–8 months
Cottonseed cake	7	2	1	12–16 months
Feather meal	13	0	0	6–9 months
Fish meal	10	5	0	6–8 months
Fish emulsion	5	1	1	<4 weeks
Soyabean meal	7	2	1	8–12 weeks
Phosphate				
Bonemeal (raw)	4	20	0	8–12 weeks
Rock phosphate	0	20	0	6–12 weeks
Potassium				
Greensand	0	1	6	6–12 months
Kelp meal	1	0.1	4	2–6 months

Source Wortman et al. [104]

bringing sub-soil nutrients to surface and leaving it there in available forms. Incorporation of grasses results in net immobilization of N in the short term, followed by mineralization in crop growth stages during tillering of wheat; the potentially mineralizable nitrogen (PAN) declines thereafter [65] and becomes negative at heading. Nitrogen is more readily available from legumes; PAN from a robust stand of legumes peaks at budding growth stage and declines slowly as reproductive growth continues. To maximize PAN, cereal cover crops are cut early for incorporation, but legumes are cut at early bud stage. Combined with organic amendments to meet plant phosphorus, potassium, and micronutrient demand, cover crops (cereals, legumes or mixtures) can be sustainable and cost-effective nutrient source for organic production.

Catch crops or short-duration cereal and legume catch crops are grown between main crops during the crop rotation and can mop up residual nitrate and other available nutrients of previous crop in rotation. Nitrogen release from N fixing leguminous tissue can be retarded by tannins and affect N need of the crops particularly in temperate climes, it will usually result in temporary immobilization of N which will be released a few months later. It is crucial that the timing of the incorporation of the catch crop materials is adjusted to the needs of the crop (to achieve synchrony). Green manures are normally allowed to decompose for 15 days before the soils are flooded and rice is transplanted, however considerable N can be lost through ammonia volatilization in this process [73]. Ghai et al. [31] in field experiments on green manuring in North India showed that transplanting rice within 5 days after turning in, is best to get high yields, N uptake and N use efficiency as the growing rice plants constantly remove the mineralized nitrogen and this tight coupling between

Table 2 Influence of decomposition period (days) of *Sesbania* spp. green manure[a] on N availability to flooded rice in alkaline soil in North India (from Ghai et al. [31])

Species	Decomp. (days)	Fresh biomass turned in ($Mg\ ha^{-1}$)	N turned in ($kg\ ha^{-1}$)	Grain yield ($Mg\ ha^{-1}$)	Total crop N uptake ($kg\ ha^{-1}$)	Fertilizer N equivalence ($kg\ ha^{-1}$)
S. cannabina	15	24.0	110[b]	5.9	67[b]	65
	10	25.0	125	6.1	86	114
	5	25.2	109[b]	6.1	89	122
	0	24.6	120	5.4[b]	84	109
Sesbania sp. PL-Se 17	15	20.9	98[b]	5.2	61[b]	50
	10	20.6	81[b]	5.1	73	81
	5	21.4	108	5.0	74	83
	0	20.1	103	5.4[b]	72	78

[a]Green manure incorporated at 45d growth
[b]Significantly different from other treatments in same species ($p = 0.05$)

the mineralization and crop uptake reduces the yield losses. Incorporation of 45d old *Sesbania cannabina* green manure containing 110 kg N could provide rice yields equivalent to those obtained with 120 kg N/ha fertilizer application (Table 2, [31]). Thus, the N of green manure can be utilized with the same or even greater efficiency as urea.

Nitrogen fixation by legume cover crops like hairy vetch and field pea, can provide up to 150 kg N/ha and can be especially useful for balancing the phosphorus and potassium supplied through manure or compost applications. Dey and Jain [20] found that N balance after rice in a rice-wheat cropping system was positively correlated with the N content but negatively correlated with lignin content and C:N ratio of the green manures. In agro-forestry, the catch crops are multi-purpose short duration perennials whose green loppings are used as fodder or directly incorporated into the soil as green manure.

Choice of catch crop has to be done carefully; if used indiscriminately, very deep roots of catch crops can deplete the subsoil so effectively as to adversely affect the deep-rooted main crop grown in rotation. The use of catch crops will usually benefit weed control. Rotational benefits of legumes (e.g., soybean, beans, alfalfa or clover) in the crop rotation can result in a nitrogen credit for subsequent crops due to biological nitrogen fixation and less depletion of soil nitrogen compared to a non-legume as the preceding crop. The nitrogen credit for forage legumes is greater than for grain or vegetable legumes because most of the nitrogen (whether scavenged from the soil or biologically fixed) is removed during harvest of the grain. Deep-rooted crops like alfalfa and *Sesbania* can scavenge immobile nutrient like phosphorus and leached nutrient like nitrate-nitrogen from deep soil layers, which may be released near the soil surface with decomposition of the crop residue for availability to subsequent crops in rotation. In North India, at 100 days growth green stem and leaf production of *Sesbania cannabina* was 21.5 and 9.4 Mg ha^{-1}; Biofertilizer value in leaf and upper tender stems was 125, 5, 81 and 12 kg ha^{-1} of N, P, K and S ha^{-1} respectively [75]. Crops like Pigeonpea (*Cajanus cajan*) produce significant amounts of litter; Rao and Gill [74] recorded 1.9 Mg ha^{-1} of litter fall in a season that contains 39.5, 2.1, 7.3 and 2.1 kg of N, P, K and S ha^{-1} respectively, that improves yields of wheat grown in rotation. The rotational benefit of legumes and N credit for succeeding cereal crops are widely known. Long term experiments over 8 years in a Vertisol showed a fertilizer nitrogen credit of soybean of 30 kg ha^{-1} in wheat [76].

Natural minerals or materials rated very slow (rock phosphate, granite meal, sand) to medium (kaolinite) in nutrient availability may be used to maintain a given level and nutrient balance in the soil. Where a rapid change in nutrient levels or balance is necessary, materials having medium to rapid nutrient availability such as gypsum and pyrites should be used. It is beneficial to add raw rock phosphate to manures when making composts rather than as a soil application. Microbial activity and localized acidity in manures and composts can increase the availability of the phosphate from rock phosphate. Generally, farmyard manure produced by farmers in India contains very low P and S. And P deficiency is very common in majority of the Indian soils and most of the soils of oilseed growing areas are also deficient in S. Therefore, the deficiency of P and S are expected in organic farms and rock phosphate is a good source of P for such soils. Naturally occurring gypsum is a good source of S in organic farming.

2.2 *Nutrient Budgets and Management*

2.2.1 Nutrient Budgeting

Nutrient budgeting is a useful balancing approach that helps to ensure that organic farmers do not unsustainably mine inherent soil nutrient reserves and at the same time do not also build them up to levels that pose risk of environmental contamination. Critical levels for soil nutrient availability can be used to make annual decisions about nutrient source and application rates; however, developing a long-term nutrient management plan and proactive management of soil fertility is advised for organic farms. Soil sampling every four years is advised to test for soil organic matter, pH, and availabilities of P, K and Zn as well as other properties of interest. Plant tissue analysis for every three to five years to diagnose potential nutrient deficiencies of perennial crops, such as fruit tree crops, is advised. A long-term accounting of phosphorus addition via manure (or other fertilizer source) and removal via harvested crops in Vertisols of Central India is shown in Table 3 [77].

Nutrient management plan describes the amount, source, placement, form and timing of the application of nutrients and soil amendments, to meet crop nutrient demand while protecting water quality and improving soil health. This takes into account the crop's nutrient needs as well as all credits in the system such as N contribution of cover crops, the mineralization of N from previous additions of

Table 3 Apparent phosphorus recovery (APR) of fertilizer P in soybean and wheat (mean of 5 years) [77]

Manure dose (t/ha)[b]	Rate of fertilizer P (kg P/ha)							
	0	11	22	44	0	11	22	44
	APR by soybean (%)				APR by wheat (%)			
0 (0)[a]	–	36.6	30.0	19.0	–	35.1	27.8	17.7
4 (5.6)	–	44.1	31.8	19.1	–	33.4	32.8	19.2
8 (11.2)	–	46.5	35.2	22.7	–	35.9	33.9	23.1
16 (22.4)	–	47.0	35.8	23.2	–	40.3	38.5	26.9
Mean	–	43.6	33.2	21.0	–	36.2	33.2	21.6

[a]Figures in parenthesis are P applied through manure
[b]Manure with fertilizer P applied to soybean whereas wheat received only fertilizer P

manures/minerals and any nitrate that may be contained in irrigation waters. The manure application rate is determined after calculating the target nutrient application rate (difference between crop need and the nutrient credits) and evaluating amounts likely to be made available by current mineralization and that lost by leaching and runoff. However, as soil fertility management in organic systems is on a longer term compared to conventional, soil analysis and interpretation has to be adapted to reflect not just the nutrient availability from some defined fractions but organic matter turnover rates, biological activity etc. It has to be different from conventional soil analysis and include indicators of soil health and quality [101].

2.2.2 Nutrient Availability

The timing, application and incorporation of organic sources is most important to achieve optimum nutrient availability and efficiency. Not all of the nutrients in manure are released in the first year of application. Nutrient carry-over from crop residues and manured soils also contributes to the fertility of the soil. Only half the N and P are considered available in the first year of application, but all of the K should be available in the first year. Surface broadcast manures can lose one half and two thirds of the available N through volatilization. Organic manures contain valuable quantities of nitrogen, phosphate and potash, experiments studying the effect of timing suggest that, in order to decrease nitrate leaching, applications of manures which contain much available nitrogen should not be made during the period September to December on freely draining grassland and arable soils [89]. Spring top dressings of dilute pig or cattle slurries and poultry manures to growing cereal crops are generally more efficient than autumn applications, particularly on freely draining soils. In reduced till systems, manures are generally best applied just before soil tillage. N mineralized from manures during the rainy season is lost by leaching or runoff, so dose is minimized or manures are mixed with high C:N ratio cereal straws to immobilize N prior to planting to reduce the risk of immobilization and release N slowly. Farmers among poor and marginal indigenous groups of south Asia and

Africa have experimented with innovative agricultural practices for centuries. In one such practice in eastern India for direct seeded rice, cow-dung is powdered and mixed thoroughly with the soil after broadcasting of rice seeds and then land surface is levelled with a wood log. This helps in hastening nitrogen mineralization from the dung to meet the plant demand by the young plants due to improvement in water holding capacity of the soil [21].

Generally, when manure is applied at rates adequate to meet the N requirements of crops, P and K are more than adequately supplied as well. But that is not always the case as the manure may not be available in sufficient quantities for application, in many situations, or the manure may not have adequate P content. So, organic manures alone may not be able to supply sufficient amount of P for optimum crop growth. The organic manures are known to decrease P adsorption/fixation and enhance P availability in P-fixing soils. Organic anions formed during the decomposition of organic inputs can compete with P for the same sorption sites and thereby increase P availability in soil and improve utilization by crops. Reddy et al. [77] observed higher apparent P recovery by soybean-wheat system on Vertisol with a combination of fertilizer P and manure (Table 3). On acid soils, as are common in tropical areas, addition of manure or easily degradable plant material to the soil also increases the availability of P to plants through competition between organic anions and phosphate anions at ligand exchange sites.

2.2.3 Synchronization of Nutrient Release

To achieve plant adapted fertilization in organic farming, it is most important to match nutrient release pattern from organics with nutrient uptake pattern of crops to achieve as closed a system as possible and nutrients are not lost by leaching or volatilization. One approach is to manipulate plant demand by controlling planting date and duration of crop growth, another is to control the quality and quantity of organic inputs applied, as for example by applying a material that immobilizes N during fallow or early crop growth, and then releases it later when crop demand is greater [81]. Nitrogen release from good quality materials (low C:N ratio of 12–14:1) is very fast,in green manures, about 53, 73 and 92% of the total N was mineralized after 4, 8 and 16 weeks of laboratory incubation in a Vertisol. From poultry manure, the corresponding values were 43%, 63% and 84%, respectively. Addition of intermediate C:N ratio (30–35) materials such as farmyard manure, soybean and chickpea straw resulted in immobilization of N during the first 6 weeks followed by release of mineral N from 8 to 16 weeks of incubation. Addition of material with wide C:N ratio (80:1) such as wheat straw resulted in a marked decrease in mineral N due to its immobilization. The total amount of N mineralized ranged from 95% in *Gliricidia* to ~28% from added wheat residues. Such information is useful to decide on combinations of different proportions of organic materials of different quality to be applied so as to achieve proper synchrony of N release with crop demand. The location of released nutrients with respect to plant roots is also part of the synchrony concept.

2.3 Organics Influence on Soil Health

The earliest rigorous comparisons showed evidence that differences between organic agriculture and conventional agriculture cannot be explained by summing up single ecological measures of biological activity. In long term field trials of over two decades in Switzerland [55], organic (ORG) treatments differed fundamentally from conventional ones. The metabolic quotient for CO_2 or qCO_2 were highest in mineral fertilized soil, whilst the organic systems, which had received organic manure, showed lower values (indicates that the microorganisms of manured soils need less energy for maintenance). Lori et al. [53] conducted a meta-analysis of 56 studies from different climatic zones and experimental duration ranging from 3 to 100+ years and concluded that organic farming enhances total microbial abundance and activity in agricultural soils on a global scale. Organic systems had 32–84% greater microbial biomass carbon, microbial biomass nitrogen, total phospholipid fatty-acids, and dehydrogenase, urease and protease activities than conventional systems. Exclusively, the metabolic quotient as an indicator for stresses on microbial communities remained unaffected by the farming systems. Categorical subgroup analysis revealed that crop rotation, the inclusion of legumes in the crop rotation and organic inputs are important farming practices affecting soil microbial community size and activity.

In arid soils, organic biodynamic fertilization improved the soil physical properties, organic carbon and microbiological properties—actinomycetes population, 16S rRNA content, glomalin content and soil enzymes significantly [4]. The culture independent analysis of eubacterial 16S rRNA gene diversity showed that organically cropped farms and orchards had more diverse bacterial community compared to the conventional. Actinobacteria were higher by 10% in organic cropping. Soil biological quality was thus improved through an alteration of the microbial community structure and function by organic amendments which selectively modify the environment and make soil ecosystems more sustainable and can hence be classified as 'ecosystems engineers' [4].

3 Precision Nutrient Management

Nitrogen fertilization is of crucial importance in obtaining optimum yield, but it is never easy to determine the optimal dose due to complex soil-plant-climate interactions and variable management. The development and implementation of site-specific nutrient recommendations have been made possible by combining the Global Positioning System (GPS) and geographic information systems (GIS). Various remote sensors could be used for this purpose such as satellite imagery, ground-based optical sensors, ground-based reflective sensors, aerial imagery, and leaf chlorophyll sensors [36] combined with soil and tissue testing [83]. Over-fertilization as a precautionary principle leads to additional unnecessary costs and environmental problems of water pollution and climate change-nitrate leaching, ammonia volatilization and denitrification. N recommendation based on soil analysis that takes into consideration of soil nitrate or any other available N sources reduces the amount of needed N and improves efficiency.

Precision agriculture has been especially helpful in tackling fertilizer loss, a major concern for farmers, as nitrogen fertilizer is one of their biggest expenses. Recent results on variable N application to wheat in Yaqui Valley, Mexico [57] show that N fertilizer rate reductions from rates (300 kg N ha^{-1}) to the economic optimum rates (123 and 145 kg N ha^{-1}) can significantly improve N use efficiency, from 36–39% at higher rates to 50–57% at optimum rates and also reduce N_2O emissions (avoiding the equivalent to 0.5–0.8 Mg CO_2e ha^{-1} yr^{-1} or, regionally, 84–138 Gg CO_2e yr^{-1}) without compromising yields. Basso et al. [7] analyzed eight years' worth of satellite data that covered corn- and soybean fields spread across 30 million ha of Midwestern states of USA and showed that only about 50% of the cropland is high-yielding, the remaining half is either of consistently low-yielding fields (26%) or those that produce a mix of low and high yields (28%). So, only half the nitrogen in the fertilizer that had been applied was actually taken up by the plants and the rest 50% lay wasted in the soil, or lost as greenhouse gases or swept up by rainfall into waterways. They concluded that streamlining the fertilizer usage can save US corn and soybean farmers an estimated $500 million in fertilizer costs, and stop 6.8 million metric tons of carbon dioxide equivalent escaping from unused nitrogen fertilizer into the atmosphere. Availability of several soil-crop simulation models paved the way for effective N management and assessment of NUE and N losses. These models integrate the effect of soil, weather, cultivar, pest, and other management practices on the growth and yield of crop [98].

3.1 Soil Based Approaches

Proper management of fertilizers is aimed at increasing the system productivity, farmer profitability, enhanced environmental protection and improved sustainability. Other agronomic and conservation practices, such as no-till farming and the use of cover crops, play a valuable role in supporting best nutrient management practices.

3.1.1 Soil Testing and Crop Response

In conventional soil testing programs, a set of critical values that characterize the estimates of soil available nutrients as low, medium or high serve as an empirical guide to rates of fertilizer application for various crops and are used by soil testing laboratories in India. The generalized recommendations of fertilizer use are applicable to the medium category of soil testing estimates with an arbitrary adjustment (decrease or increase by 25–50%) for high and low categories of soil test estimates. This classical approach relies only on the soil analysis and does not take into account yield target or residue management in previous crop etc. Some workers also reported the lack of correlation between soil-test based data for available nutrients and crop yield [24] or the approaches for developing integrated nutrient recommendations from soil test data [95]. Due to this, nutrient management in developing countries like India is characterised by abuse of nitrogen, disuse of potassium, coupled with overuse of phosphorus. These outmoded blanket prescriptions are the chief reason for imbalanced plant nutrition. Farmers own perception compounds the problem further and leads to over application in many cases, for example the overuse of zinc in sodic soils of Punjab and Haryana in north India.

The Indian Council of Agricultural Research's Project on Soil Test Crop Response (STCR) uses multiple regression approach to develop relationships between the levels of soil nutrients determined in the laboratory test and crop response to fertilizers in the field that permits balanced fertilization through right kind and amount of fertilizers to achieve a given yield target. It is unique because it predicts the level of yield the farmer can hope to achieve if good agronomic practices are followed in raising the crop. The resultant fertilizer adjustment equations have been tested in follow up and frontline demonstrations conducted in different parts of the country. This targeted yield concept strikes a balance between 'fertilizing the crop' and 'fertilizing the soil' and provides a scientific basis for balanced fertilization and strikes a balance between applied nutrients and soil available nutrients. In the STCR targeted yield approach, it is assumed that there is a linear relationship between grain yield and nutrient uptake by the crop, and for obtaining a particular yield, a definite amount of nutrients are taken up by the plant. Once this requirement is known for a given yield level, the fertilizer needed can be estimated taking into consideration the contribution from soil available nutrients. Several thousand demonstrations have been conducted all over India that attest to the superiority of STCR-based fertilizer recommendation for different crops over blanket recommendation and yield targets within ±10% variations of the target set were achieved along with better response ratios (Table 4) over a wide range of agro-ecological regions [18]. Ramamurthy et al. [71] concluded that due to inherent soil and management variation within an agro-climatic zone,

Table 4 Agronomic use efficiency of nutrients (kg grain/kg nutrient applied) in various crops in field demonstrations in farmers fields across 10 states of India [18]

Crop	No. of trials	Farmer's practice	STCR-IPNS recommended practice
Rice	120	11.4	16.8
Wheat	150	10.3	14.2
Maize	35	12.7	17.7
Indian mustard	45	8.0	8.2
Rapeseed	25	4.8	7.6
Groundnut	50	5.1	6.8
Soybean	17	9.6	12.2
Chickpea	35	6.1	9.4

broad recommendations for fertilizer usage as for example sugarcane in Karnataka, India could lead to insufficient nutrient application. With the availability of soil fertility maps at watershed and mandal and district level, fertilizer recommendations based on such maps can form the basis of precision farming.

In another subset, site-specific nutrient management (SSNM) aims at optimizing agriculture production by managing both the crop and the soil to improve yields and nutrient uptake. This was developed for irrigated rice systems for Asia by IRRI in 1996 [23] and subsequently tested in more than 200 irrigated rice farms across Asia. It focused on managing field-specific spatial variation in indigenous NPK supply, temporal variability in plant N status occurring within a growing season and medium-term changes in soil P and K supply resulting from actual nutrient balance. The approach required a data management option to predict soil nutrient supply and plant uptake in absolute terms in the high-yielding irrigated rice systems in Asia. A modified QUEFTS model [41, 103] was used for this purpose. It described the relationship between grain yield and nutrient accumulation as a function of climatic yield potential and the supply of the three macronutrients. In a situation of balanced nutrition, the QUEFTS model assumed a linear relationship between grain yield and plant nutrient uptake at constant internal efficiencies until yield targets reach about 70–80% of yield potential. As yields approach the potential yield, the internal nutrient efficiencies decline as the relationship between grain yield and nutrient uptake enters a non-linear phase. To model this in a generic sense required the empirical determination of two boundary lines describing the minimum and maximum internal efficiencies of N, P and K in the plant across a wide range of yields and nutrient status. A database containing more than 2000 entries on the relationship between rice grain yield and nutrient uptake was used to derive the generic boundary lines of internal efficiencies [22]. The balanced N, P and K uptake requirements for 1000 kg of rice grain yield were estimated from the respective envelope functions as 14.7 kg N, 2.6 kg P and 14.5 kg K, which is valid for the linear phase of the relationship between yield and nutrient uptake. The corresponding borderlines for describing the minimum and maximum internal efficiencies were estimated at 42 and 96 kg grain kg^{-1} N, 206 and 622 kg grain kg^{-1} P and 36 and 115 kg grain kg^{-1} K, respectively [103]. The parameters were found to be valid for any site in Asia at which modern rice varieties with a harvest index of about 0.45–0.55 were grown. A study of 179 rice farms in 6 Asian countries found that SSNM led to yield increases of 7% and total profitability increases of 12% [24].

Various fertilizer decision support tools to implement 4R nutrient stewardship (right source, right rate, right time and right place) like *Nutrient Expert* combined with ‘Green Seeker’ to guide fertilization improved the N use efficiency and reduced the environmental footprint of wheat production under no-till in North-west India [82]. In North-central China [106] fertilizer recommendation from *Nutrient Expert* for summer maize improved the yields, reduced N fertilizer by 35%, improved the agronomic efficiency of N and reduced N_2O and total GHG emissions by 17–18% over soil test-based fertilizer recommendations.

3.1.2 Fertigation

Fertigation is an important part of precision farming, in which dissolved fertilizers are delivered to the crop through the irrigation system which allows the application of precise rates of water and fertilizers at the root zone, accurately timed to the crop needs [56], a technique first started in the late 1960s in Israel [90]. Concentrated fertilizer solutions are prepared in stock tanks and injected into the irrigation water, using fertilizer injectors. The irrigation system can be drip irrigation, sprinkler system, pivot or different configurations of hydroponic systems. Since it involves frequent application of small quantities of water on or below the soil surface as drops, irrigating the root zone instead of the entire land, and matching with the consumptive water demand of plant, the water and nutrient use efficiency are both improved resulting in savings and fertilizers, minimises leaching and allows flexibility in timing of fertilizer application and reduces the labour required for applying the fertilizers and pesticides [47]. But most of the systems do not yet have the capability of using feedback from sensors installed in the field and adjust fertilizer rates accordingly. Some drawbacks like high cost of setting up and precipitation can be minimized by greater availability of soluble and compatible fertilizers, good quality water, and synchronization of plant demand and supply of nutrients, lack of precipitation of phosphatic fertilizers with micronutrient mixtures, availability of corrosive resistant fertigation systems.

Sub-surface drip irrigation and fertigation in maize-wheat under conservation agriculture in Punjab, India saved 50–60% irrigation water and 25% fertilizer nitrogen for each crop. Similarly, the nitrogen use efficiency was higher in rice and wheat [16]. Fertigation results in optimum growth, improved yield and produce quality [44, 84] besides substantial water saving to the tune of nearly 12–84% in different crops [64]. Fertigation improves nutrient use efficiency (Table 5) and minimizes the pollution of ground water by preventing losses through run off and leaching. Fertigation has been found to save 20–30% fertilizers in Banana besides improving the yield and quality as compared to the conventional fertilizer application [62]. Teixeira et al. [97] reported 36% higher N and K use efficiency. Nanda [63] obtained 60–70% higher yield through drip fertigation with water soluble fertilizers (WSF) in banana as compared to soil application of fertilizers Using targeted yield model, fertilizer prescriptions under STCR-IPNS developed for transgenic cotton on Periyanaickenpalayam soil series (Vertic Ustropept) under drip fertigation was found useful for Tamil Nadu [68].

Table 5 Fertiliser use efficiency (%) in fertigation [91]

Nutrient	Soil application	Drip + soil application	Drip + fertigation
N	30–50	65	95
P_2O_5	20	30	45
K_2O	60	60	80

3.2 Plant Based Approaches

3.2.1 Foliar Fertilization

Foliar fertilization offers a specific advantage over soil fertilization when plant demand for nutrients exceeds the capacity of the roots to absorb nutrient and when environmental conditions limit the effectiveness or prevent the application of nutrients to the soil. Further, soils with high levels of aluminum, calcium, and iron reacts with soil applied fertilizers to form insoluble compounds which limit the effectiveness of soil-applied phosphorus (P) fertilisation strategies [38]. Although not commonly used in dryland cropping systems to date, foliar P fertilisation may allow a tactical response to prevailing seasonal climatic conditions, with the added benefit of reduced input costs at sowing. The current knowledge about the factors that influence the ultimate efficacy of foliar applications are incomplete, however the following key principles are recognized [66] (a) physiological age of crop—a decreasing response with increasing crop age, (b) tissue test status—a crop that is very deficient will have limited potential to overcome deficiency at low foliar application rates, while a crop that is sufficient will not be responsive, and (c) Leaf area of crop: sets foliar nutrient uptake potential and depending on physiological age of crop sets photosynthetic potential for grain filling.

The response of various soybean cultivars to foliar N, P, K, S supply has been encouraging [12] when applied during the seed-filling period. Foliar N supply to soybean was found to be an effective means to replenish N in the leaves and resulted in higher yields in contrast to soil fertilization alone [29]. In wheat (*Triticum* sp.), several studies have shown that foliar sprays of N increased grain protein. While studying mixed foliar sprays, Alston [1] attributed the yield response of wheat to foliar application of NP to be the result of N rather than the P in the spray. Optimum timing for N sprays on wheat showed that post-pollination foliar N gave the highest grain protein [10, 11, 32]. The benefits of late season foliar applications are influenced by both cultivar and plant N status [99]. Foliar applications of P are the most effective way to supply P to crops late in-season whenever there is a need [33]. The effectiveness of foliar P application in cereals has been reported in many investigations [6, 13, 33, 59].

Foliar fertilizers can be used to enhance crop quality both in terms of grain protein and Zn content [14, 25]. Nitrogen foliar fertilisation can result in increased levels of amino acids in the young wine and with or without sulphur foliar nutrition can enhance the aromatic expression in wines [45]. Rabani-Foroutagheh et al. [70] found that split foliar fertilisation and the fertiliser type significantly increased saffron yield, number of flowers and crocins (colour), whereas it decreased the picrocrocin (taste) and safranal (odour) content of the saffron.

3.2.2 Plant Improvement

Knowledge on improving nitrogen use efficiency and its components and the key molecular players in N uptake, assimilation, and utilization has grown significantly [43, 96] however, improving the efficiency of these processes has proved difficult. Improving the crop N use-efficiency (NUE) requires in-depth understanding of unknown co-regulatory mechanisms integrating growth, N assimilation and carbon (C) fixation. Li et al. [48] showed that balanced opposing activities and physical interactions of the rice Growth-Regulating Factor 4 (OsGRF4) transcription factor and the DELLA growth-inhibitor confer homeostatic co-regulation of growth, C and N metabolism. OsGRF4 promotes and integrates N assimilation, C fixation and growth, whilst DELLA inhibits them. The DELLA proteins are a family of putative transcriptional regulators that inhibit the cell proliferation and expansion that drives the growth of plant organs. In consequence, the DELLA accumulation characteristic of the green revolution varieties (GRVs) confers yield-enhancing dwarfism, but also reduces NUE. Plant growth is stimulated via destruction of DELLA proteins in the 26S proteasome. However, GRV NUE and grain yield are increased by tipping the OsGRF4-DELLA balance towards increased OsGRF4 abundance. Modulation of plant growth and metabolic co-regulation thus enables novel breeding strategies for future sustainable food security and a new Green Revolution.

Swarbreck et al. [93] have provided a roadmap to integrate the regulation of nitrogen uptake and assimilation into varietal selection and crop breeding programs. The overall goal is to reduce nitrogen inputs by farmers growing crops in contrasting cropping systems around the world, while sustaining yields and reducing greenhouse gas (GHG) emissions. Considering N responsiveness and its underlying mechanistic regulation, as well as the introduction of varietal or advance breeding line screening varieties for low N requirement and highly responsive varieties in crop breeding pipelines, is likely to provide new means to develop new varieties with a low economic N optimum.

3.2.3 Leaf Colour

Site specific N management studies using chlorophyll meter, have shown that N savings can be effected along with improved plant N use efficiency [88]; their results on wheat recommended application of 30 kg N ha^{-1} as basal dose, 45 kg N ha^{-1} at crown root initiation and 30 or 45 kg N ha^{-1} at maximum tillering depending on leaf color as opposed to blanket dose of 120 kg N ha^{-1} in two equal splits at planting and CRI. Simple leaf color charts (LCC) have been developed for rice and wheat [40, 85, 100] as a useful low-cost tool to support decision making on the timing and quantity of N fertilizer application. The graded panels with differing shades of green are compared with the colour of the adaxial leaf surface. It is of low cost and easy to use to assess the N requirement of the crop. The dissemination of this low-cost tool to a model village in Punjab has led to a reduction in N application of an average of 75 kg N ha^{-1} in rice and 50 kg N ha^{-1} in wheat. However, the use of leaf color chart may not be uniformly applicable to all varieties that differ in inherent leaf color and regions that differ in climate, thereby necessitating individual or group standardization in different cultivated areas. In the Soil Test-Crop Response (STCR) methodology of site-specific nutrient management, recent development of algorithms of leaf colour chart, SPAD and fieldscout CM 1000 m values at three critical growth stages with yield in rice-wheat system showed promising results [19].

3.3 Product Based Approaches

3.3.1 Slow Release and Customized Fertilizers

Much of the nitrogen applied by farmers goes into the atmosphere as ammonia and to a lesser extent as oxides of nitrogen. Phosphorus ends up in watersheds creating ideal conditions for eutrophication. Installing precision systems based on autonomous vehicles and sensors is costly, so only large-scale operations tend to have them. In comparison, advanced controlled-release fertilizers are relatively inexpensive and can ensure that significantly higher levels of nutrients reach the crops, leading to higher yields with less fertilizer application. Slow-release fertilizers typically consist of granules or tiny capsules filled with nutrients; they need to meet each of three controlled release criteria, viz., not more than 15% release in 24 h, not more than 75% release in 4 weeks and at least about 75% release at the stated release time. The most popular are the condensation polymers-urea formaldehydes, urea isobutyraldehydes, urea crotonaldehydes. These are water insoluble and N is slowly hydrolysed from the polymer chain. The others are the coated ones like S-coated, polymer coated and latex coated urea products. Here N is slowly released by diffusion through the coatings. Gel based materials contain urea absorbed within acrylic gels for example. Urease inhibitors (NTBT) and nitrification inhibitor incorporated urea materials (N-Serve, DCD) are known for long. Sophisticated materials and manufacturing techniques have been developed that can tune the outer shells and alter nutrient-release rates in desired ways as the soil's temperature, acidity or moisture changes. The global consumption of controlled- and slow release fertilizers has increased at an average annual rate of 30% during 2009–14. Polymer-coated urea is now one of the most widely used controlled release products in arable cropping systems [105].

Although controlled-release technologies are more efficient, they do not eliminate all the drawbacks of fertilizer use. The products still include ammonia, urea and potash, for example; producing these substances is energy-intensive and contributes to greenhouse gas production and climate change. However, the losses are mitigated. They are more expensive, however sulphur coated urea is the least expensive. In a field evaluation of coated and modified urea products for submerged rice in a sandy loam soil in North India [72] the peak concentrations of ammoniacal N (AN) in the floodwater (pH 9.5–10.0) were >12 g m^{-3} and ammonia volatilization losses were ~54% of applied N when 120 kg N ha^{-1} were applied as a whole dose U(W). When applied in 3 equally split doses U(S), peak AN levels reduced to ~5 g m^{-3}

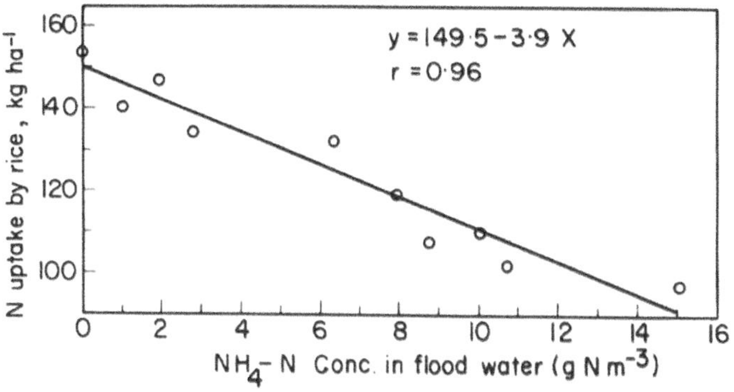

Fig. 1 Relationship between nitrogen uptake by rice at harvest and nitrogen concentration in flood water two days after fertilization and rice seedlings transplantation (Rao [72])

and losses to 45%. Sulphur coasted urea (SCU) and urea super granules (USG, deep placed) were effective in reducing peak AN levels to <2 g m^{-3} and N losses to 17 and 3% respectively. N uptake efficiency with U(S) was 41%. SCU was ineffective in improving grain yields but improved N recovery to 58%. USG increased grain yields only in first year by 20% over U(S) and improved N uptake to 61%. A negative linear relationship was observed between N uptake by rice at harvest and AN levels in floodwater two days after fertilization (r = 0.96, Fig. 1) which was recommended as a simple predictable index to evaluate the use efficiency of applied N fertilizers. Based on a meta-analysis of 32 studies in China, Linquist et al. [50] reported increase of 5.7 and 8.0% in rice yield and N uptake when Enhanced Efficiency Nitrogen Fertilizers (EENF) (i.e. nitrification and urease inhibitors, neem, and slow release fertilizers) were applied. In Bangladesh, deep placement of urea super granules (USG) in rice was widely popularized and led to 30% reduction in N fertilizer over prilled urea [69].

The fertilizers, which are able to cater regional soil-crop-stage specific fertilizer requirements, can be called "Customized Fertilizers". Customized fertilizers facilitate the application of the complete range of plant nutrients in the right proportion and to suit the specific requirements of a crop in different stages of growth, and are more relevant under site specific nutrient management practices. Fortified fertilizers are generally common fertilizers to which one or more specific nutrients have been added in order to increase their nutrient content and make them more versatile. Three important fortified products included in the Fertilizer Control Order of Government of India are Zincated urea—2% Zn, Boronated SSP—0.18% B, and 10:26:26 fortified with 0.5% Zn and 0.3% Boron [27]. The Government of India has introduced a mandatory provision to coat all urea fertilizers with '*neem*' oil, a nitrification inhibitor.

3.3.2 Nano-Fertilizers

Nanotechnologies refer to those technologies that contain at least 50% natural or manufactured particles in the size range from 1 to 100 nm. Nano-based smart fertilizer delivery systems have the ability to provide more efficient and targeted delivery to specific plant cells due to their size-related high surface area. Also, they show enhanced stability in the environment, and slow or controlled release which improves the availability of nutrients to crops and improves the stress resistance of crops. The increased use efficiency can reduce the consumption of fertilizers, besides reducing environmental fallouts. Nano-biosensors can enhance this process even further by enabling smart delivery systems to precisely release nutrients in response to environmental triggers such as heat and moisture and nutrient demand signals like carbonaceous root exudates, thus enabling real-time monitoring and control.

Liu et al. [52] have shown that nano-composites containing organic polymer intercalated in the layers of kaolinite clays can be used as a cementing material to regulate the release of nutrients from conventional fertilizers. A patented nano-composite consisting of N, P, K, micronutrients, mannose and amino acids was shown to increase the uptake and utilization of nutrients by grain crops [42]. Bansiwal et al. [5] developed a surface modified zeolite as a carrier of slow release phosphatic fertilizer. Mukhopadhyay [60] developed patented nano-fertilizers by intercalating phosphate ions in kaolin clay mineral and zinc-based nanomaterials by using clay receptacles and embedding into a polymer matrix. Nano-fertilizers are capable of releasing nutrients, especially NO_3-N for more than 50 days while nutrient release from conventional fertilizer (urea) ceased beyond 10–12 days [92]. Application of nano-P increased the growth rate and seed yield of soybean by 33 and 20% compared to regular P fertilizer Ca $[(H_2PO_4)_2]$ [51]. However, these fertilizer materials need to be screened for bio-hazards before large-scale adoption under field conditions.

3.4 Fertilizers Influence on Soil Health

Soil acidification following heavy nitrogen application is well known. Guo et al. [34] reported significant soil pH reductions in large production areas in China; N cycling released 20–221 kilomoles of hydrogen ion (H^+) ha^{-1} yr^{-1} and base cations uptake contributed a further 15–20 kilomoles of H^+ ha^{-1} yr^{-1} to soil acidification in four widespread cropping systems. However, application of recommended quantities of fertilizers are beneficial to soil productivity. Ladha et al. [46] analyzed data following time by fertilizer N response ratio in 114 long-term experiments and found declines of 7–16% in organic C and 7–11% in organic N with no N amendments. But in soils receiving fertilizer N, average increase of 8% in organic C and 12% in organic N were reported. Addition of organic matter (i.e., manure) increased SOM, on average, by 37%. Their results showed that in addition to its role in improving crop productivity, synthetic fertilizer N significantly reduces the rate at which SOM is declining in agricultural soils, worldwide. In long-term fertilization trials in cropping systems worldwide, a meta-analysis of 107 datasets from 64 trials [30] showed that mineral fertilizer application led to a 13% increase in organic carbon and 15% increase in the microbial biomass (Cmic) above levels in unfertilized control. While N fertilization tended to reduce Cmic in soils with a pH below 5 (−11%), it had a significantly positive effect in soils of pH ranging from 5 to 7 (+8%) and pH 7 or higher (+48%).The input of N *per se* does not seem to negatively affect Cmic in cropping systems. Singh [86] reviewed the positive benefits of balanced fertilization on soil organic carbon in various long-term experiments in India in various climatic zones. Benbi and Brar [8] showed that intensive agriculture with fertilizers in Punjab in North India, led to increase in soil organic carbon by 38% from 0.29 to 0.4% over a period of 25 years (1981–2006). It can be expected that when nutrients are applied in precision mode these differences will be even more and precision agriculture can also contribute to improvement of soil health besides saving nutrients and reducing losses.

4 Integrated Nutrient Management

Conventional precision farming and organic farming approaches have been well researched, contrasted and compared in many studies. In most situations in developing countries, balanced fertilization, integrated nutrient management (INM) and integrated pest management are in use by farmers for long. Addition of organic manures alone or in combination with fertilizers has been found in several studies to increase the efficiency of soil, manure and fertilizer nutrients as compared to conventional application of fertilizers [77]. The basic principle of INM is to maintain or adjust plant nutrient supply to achieve a given level of crop production by optimizing the benefits from various sources of plant nutrients like chemical fertilizers and organic sources.

Three approaches were developed for optimizing organic and inorganic nutrient doses.

Approach 1: Substitution of certain proportion of the generally recommended dose of NPK (25–50% of GRD) through organic manures and biofertilizers. These doses for different crops developed through model agronomic trials are used. Depending upon the availability of quantity of organic manures, a certain proportion of GRD particularly N is substituted through organic manures and biofertilizers. This is a widely used approach of INM followed by the soil testing laboratories.

Approach 2: A soil test based NPK dose for different crops is computed by soil test crop response correlation equations (STCR) using target yield, initial soil test values, nutrient requirement of crops and use efficiency of soil and fertilizer nutrients. Depending upon the amount of organic manures available, a certain proportion of this STCR based dose (25–50%) is substituted through organic manures and biofertilizers.

Approach 3: STCR-IPNS based fertilizer and manure prescription equations have been developed to derive doses of NPK fertilizers and organic manures to different crops using nutrient availability coefficients of organic manures, nutrient requirement of crops and use efficiency of soil and fertilizer nutrients. If a farmer knows the soil test values of his field, target yield and amount of FYM available on his farm, he can obtain an optimum dose of NPK and organic manures from the fertilizer prescription equations. Approaches 2 and 3 were found most economical as compared to the Approach 1 in several field trials conducted at various STCR centres [2]. These approaches need to be popularized among farmers through extension agencies.

4.1 INM Under Irrigated Agriculture

Long-term studies being carried out under various long-term projects of ICAR in India [39] have shown that it is possible to substitute a part of fertilizer N needs of both summer and winter crops by farm yard manure (FYM) without any adverse effect on the total productivity of the system in major cropping systems such as rice-rice, rice-wheat, maize-wheat, sorghum-wheat, pearl millet-wheat, maize-wheat and rice-maize. Sustainable yield index (SYI) of maize-wheat cropping system after 27 years at Ranchi was the highest with use of 100% NPK and FYM ([94], Table 6). In a comprehensive study on INM in irrigated rice-wheat in Bangladesh, Haque et al. [35] applied 5 Mg ha^{-1} of cowdung (CD) or cowdung bioslurry (CDB), and 3 Mg ha^{-1} poultry manure (PM) or poultry manure bioslurry (PMB) to wheat and compared it with NPKS chemical fertilization (120: 23:70: 12 in wheat) or in INM mode where nutrients added in manures were accounted (assuming 50% availability) and rest was added as chemical fertilizers (amounted to 75–80% N) to achieve same level of total NPKS fertilization. Zinc and Boron were applied @ 2 and 1.5 kg ha^{-1}. The rice grown in rotation was fertilized uniformly 75, 8, 30, 8 kg ha^{-1} of NPKS. The total system productivity improved by application of PMB, PM, CDB and CD by 11.7, 8.9, 5.4 and 3.1% respectively over chemical fertilization. Field trials in farmers' fields in Madhya Pradesh, Central India on soybean and wheat rotation for two years at four sites, showed that integrated use of fertilizers and organics

Table 6 Effect of INM on sustainable yield index (SYI) in maize-wheat system after 27 years at Ranchi [94]

Treatment	Grain yield (t/ha) (average of 27 years)		SYI	
	Maize	Wheat	Maize	Wheat
100%NPK	0.80	1.70	0.07	0.14
100%NP	0.55	1.20	–	–
100%N	0.11	0.12	–	–
100%NPK + FYM	2.80	2.50	0.30	0.23
No fertilizer	0.50	0.76	0.06	0.09

Table 7 Mean effects of seven nutrient management treatments (T1 to T7) on soybean and wheat yields (Mg ha^{-1}) and nutrient uptake (kg ha^{-1}) in mother trials during 2006–07 at four sites in farmers' fields in Vertisols of Central India [78]

Treatments in soybean	Soybean				Wheat			
	Seed	N	P	K	Grain	N	P	K
T1: Balanced fertilization 100%NPKSZn	1.81	162	15	73	5.17	134	20	117
T2: INM 1—50%NPKSZn + FYM 5 Mg ha^{-1}	1.90	173	16	80	4.94	123	18	106
T3: INM 2—INM1 + *Rhizobium*	2.04	184	18	88	4.95	124	18	107
T4: FP—50% NPKSZn + FYM 2 Mg ha^{-1}	1.59	137	13	60	4.21	99	14	86
T5: 50%N + 75%P + 100% KSZn + *Rhizobium* + PSB	1.69	149	14	67	4.23	101	14	86
T6: FYM 5 Mg ha^{-1}	1.63	139	13	61	4.09	95	13	81
T7: FYM 5 Mg ha^{-1} + *Rhizobium* + PSB	1.53	128	12	55	3.99	90	12	77
LSD ($P = 0.05$)	0.14	12	1	9	0.35	18	3	21

Note In wheat; T1 is same; T2 = 75% of T1; T3 = T2 + PSB; T4 = 2/3rd N + full dose of P; T5 = T3 + *Azotobacter*; T6 = FYM 8 Mg ha^{-1}; T7 = FYM 8 Mg ha^{-1} + *Azotobacter* + PSB

alone (INM1) or with microbial inoculants (INM2) gave similar yields and nutrient uptake than either balanced chemical fertilization (BF) and was better than supplying nutrients solely through FYM or FYM plus microbial inoculants ([78], Table 7). In the INM option inorganic nutrient application of (kg ha^{-1}) 12 N, 13 P, 8 K, 10 S, and 3 Zn per crop of soybean and 22 N, 3 P, 4 K and 5 S were saved.

4.2 INM Under Rainfed Agriculture

Studies in the ICAR-AICRP on Dryland Agriculture (AICRPDA) centers showed that 1 kg of fertilizer N produced additional grain yield varying from 4.3 to 38 kg in a variety of crops (rice, pearl millet, summer and winter sorghum, wheat, safflower and mustard) grown under different rainfall environments and diverse soil types [80]. Typically, green manures could meet half the N requirements of a crop. Inclusion of legumes in a rotation benefitted the succeeding crop equivalent to 10–30 kg N ha^{-1}. Short duration legumes such as cowpea benefitted the succeeding crops significantly. INM in combination with legume-based crop is recommended for higher productivity. INM systems, besides nutrient supplementation, enhanced soils' ability to hold additional water and resulted in favorable soil biological interactions. Long term INM trials for more than 35 seasons at AICRPDA centers indicated that fertilizer cost can be reduced by substitution with organics. In most of the situations, the yield sustainability was higher when the recommended dose of fertilizer was applied.

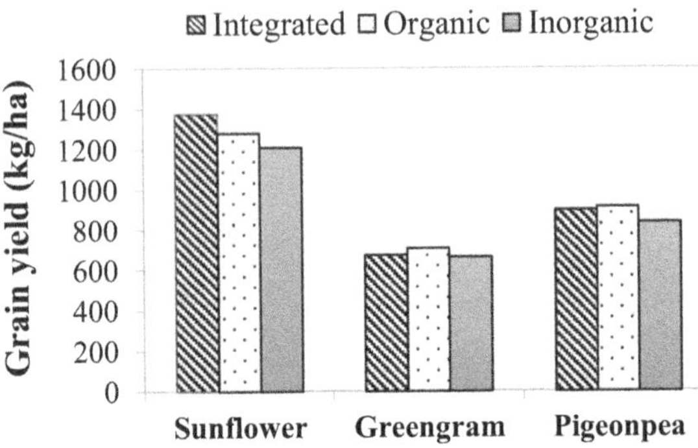

Fig. 2 Effect of integrated, organic and inorganic fertilizer treatments on grain yields of sunflower, green gram and pigeon pea [79]

In cereals, higher sustainability was obtained when the recommended dose of nutrients was applied through chemical sources. In case of oilseeds, however application of half the dose as chemical fertilizers and half through organic source led to higher sustainability values. Available nitrogen, organic carbon and phosphorus content in soil were increased with organic fertilizer application including green leafy manures. Application of crop residues in combination with chemical fertilizer resulted in higher sustainable yield and maintained higher levels of nitrogen, phosphorus and organic carbon.

Field experiments conducted during kharif 2017 showed that the seed yield of rainfed sunflower was 14% and 7% higher with Integrated Nutrient Management (INM) than that with inorganic and organic management, respectively (Fig. 2). However, plots under organic management gave marginally higher seed yield of green gram and pigeon pea under rainfed conditions. Soils under organic management had significantly higher soil organic C (0.67%), K and micronutrients as compared to inorganic production systems [79]. Similarly, in rainfed upland ecosystem of direct seeded rice in Ultisols of north-east India, application of cattle manure (10 Mg ha^{-1}) was consistently better for rice, green gram and pea (residual effect) as compared to INM options of 5 Mg ha^{-1} combined with NPK (100, 50 or 25%) in term of yield, soil organic carbon and nutrients [17].

4.3 *Improvement of Soil Health*

A number of long-term experiments have shown that integrated nutrient management systems are the best for maintaining soil health [87]. The broad conclusions are that cultivation without adding fertilizers or by adding only nitrogen degraded soil quality,organics are most superior followed by integrated nutrient management (NPK + FYM) and least in NPK only. There was greater microbial biomass, carbon sequestration and soil quality index in long term experiments with NPK + FYM fertilization as compared to NPK alone [15]. In rice-wheat in an Inceptisol, the partial substitution of fertilizer N by farmyard manure, sewage sludge and a combination (of FYM +

biofertilizer + crop residues/green manure) resulted in a high magnitude of increase in dehydrogenase activity, microbial biomass carbon and potentially mineralizable nitrogen (57%), but reduced the metabolic quotient which implied an accumulation of stable organic C under organic nutrient management [9].

5 Precision Agriculture Versus Organic Agriculture

Both precision agriculture (PA) and organic farming (OF) are solutions aimed at countering and mitigating the problem arising from chemical fertilizers usage. While both the approaches achieve this aim, PA is principally centered around improving nutrient use efficiency, fertilizer savings and reducing environmental foot print of N. But strategies are needed to ensure fertilizer use efficiency in the short-term and maintenance of soil fertility in the long term. If soil fertility is to be maintained or even increased, heavier doses of fertilizers have to be used which will be inevitably be lost, but if fertilizers are applied in low doses, native sources will be reduced and soil fertility would deplete at a faster rate. The two approaches are pulling in different directions and it is therefore necessary to adjust the fertilizer practices over seasons in such a way so as to strike a balance between the two [19].

Organic agriculture is principally aimed at improving soil and plant produce quality. OA has an in-built element of environmental safety by avoiding emissions and other concerns associated with fertilizer N usage. But OA generally lacks the potential for a short-term on-the-go reaction especially with respect to fertilizer application. Further on, in low external input sustainable organic farming, nitrogen, the key nutrient for agricultural production, is mainly provided by symbiotic legume-*Rhizobium* N_2 fixation, a process that is already related to site-specific conditions as well as small quantities of organic manure, that delivers plant-available N also strongly in accordance with site-specific conditions for mineralization. Integrated nutrient management (INM) strives for the middle ground where both the aims of PA and OF on fertilizer use efficiency, high yields and soil quality are achieved.

Organo-mineral fertilization has the advantages of both nutrient supply in sufficient quantities and efficiency associated benefits. In Europe, organic fertilizer products have to conform to new EU fertilizing products regulations (FPR) effective June 2022 [3]. New regulation will include all kinds of fertilisers such as biostimulants, industry by-products, mineral and organic fertilisers and more. For marketing as one declared source of primary nutrients it has to contain 2.5% N or 2% P or 2% for K. For more than one declared source, minimum 1% each of NPK and total 4%. An organo-mineral fertilizer shall be a co-formulation of one or more inorganic fertilisers and one or more materials containing organic carbon and nutrients of solely biological origin. It may contain peat, leonardite and lignite, but no other material which is fossilized or embedded in geological formations and shall have minimum 7% organic matter and shall not contain more than 16% N as ammonium nitrate. Minimum N, P and K content for sole sources are same as above except that of the 2.5% N (of which at least 1.0% is as organic N. And for organo-mineral NPK solid mixtures, minimum 2% each shall be required (with at least 0.5% organic N). In view of the various developments discussed in the foregoing sections, some improved nutrient management scenarios can be envisaged as below:

- Going beyond nutrient use efficiency in precision agriculture for making fertilizer recommendations which also take care of soil health aspect as given in examples in previous section.
- Improve INM by precise application of chemical fertilizers along with use of microbial inoculants e.g., *Rhizobium* inoculation of legumes and plant growth promoting bacteria for cereals.
- Since INM involves the use of chemical fertilizers, the produce cannot be termed 'Organic'. However, if nutrient sources are delivered purely in organo-mineral forms through a combination of manures and minerals or mineral-fortified enriched manures, it will be an ultimate win-win solution.
- Organo-mineral precision fertilization (OM-PF) strategies will need studies on kinetics of nutrient release in various soil types, cropping systems and climatic zones. Soil testing protocols need to be developed to assess bio-available organic nutrient pools in the long term. OM-PF has a potential to provide solutions that integrate the benefits of all fertilization strategies, while achieving improved soil health and produce quality and environmental safety.
- Breeding to produce plant varieties with improved root proliferation and root architecture as a short-term goal along with metabolic engineering to improve response of plants to applied nitrogen and nutrient use efficiency would pave the way for a reduction in fertilizer N usage.

6 Conclusions and Perspective

- It is widely believed that the benefits of precision agriculture can only be realized on large farms with huge capital investments and experience with information technologies. However, inexpensive and easy-to-use hand-held devices and techniques are being developed for wider use by all farmers.
- Conventional methods of soil testing are time-consuming. Preferably fast, non-contacting and non-destructive methods should be available to obtain the required information on soil available nutrients that can be fed easily into systems to recommend fertilizer applications.
- Assessment of the nitrogen content, biomass, and nitrogen uptake of plants by contactless optical measurements is a promising technique for site specific management decisions of farmers.
- Integrated nutrient management (INM) can be further improved by addition of the chemical fertilizers by precision management approach and microbial inoculants to improve biological nitrogen fixation, bio-available nutrients and soil health.
- Soil test-based recommendations provide a scientific basis for balanced nutrition not only between the applied nutrients themselves but also with the soil available nutrients. However, constraint analysis with respect to biological and physical

properties has seldom been researched. These two factors account for about one-third of unaccounted variations in soil health assessment.

- Organo-mineral precision fertilization (OM-PF) requires long term research on nutrient availability, crop yields, effects on soil biology and fertility.

References

1. Alston, A.M.: Effects of soil water content and foliar fertilization with nitrogen and phosphorus in late season on the yield and composition of wheat. Aust. J. Agric. Res. **30**, 577–585 (1979)
2. Annual Report: Annual Report 2008–09. ICAR-Indian Institute of Soil Science, Bhopal (2008)
3. Anonymous: L170. Off. J. Eur. Union **62**, 132 pp (2019)
4. Aparna, K., Pasha, M.A., Rao, D.L.N., Krishnaraj, P.U.: Organic amendments as ecosystem engineers: microbial, biochemical and genomic evidence of soil health improvement in a tropical arid zone field site. Ecol. Eng. **71**, 268–277 (2014)
5. Bansiwal, A.K., Rayalu, S.S., Labhasetwar, N.K., Devotta, S.: Surfactant-modified zeolite as a slow release fertilizer for phosphorous. J. Food. Chem. **54**, 4773–4779 (2006)
6. Barel, D., Black, C.A.: Foliar application of P. II. Yield responses of corn and soybeans sprayed with various condensed phosphate and P-N compounds in greenhouse and field experiments. Agron. J. **71**, 21–24 (1979)
7. Basso, B., Shuai, G., Zhang, J., Robertson, G.P.: Yield stability analysis reveals sources of large-scale nitrogen loss from the US Midwest. Sci. Rep. **9**, 1–9 (2019)
8. Benbi, D.K., Brar, J.S.: A 25-year record of carbon sequestration and soil properties in intensive agriculture. Agron. Sustain. Dev. **29**, 257–265 (2009)
9. Bhaduri, D., Purakayastha, T.J., Patra, A.K., Chakraborty, D.: Evaluating soil quality under a long-term integrated tillage–water–nutrient experiment with intensive rice–wheat rotation in a semi-arid Inceptisol, India. Environ. Monit. Assess. **186**, 2535–2547 (2014)
10. Blandino, M., Reyneri, A.: Effect of fungicide and foliar fertilizer application to winter wheat at anthesis on flag leaf senescence, grain yield, flour bread-making quality and DON contamination. Eur. J. Agron. **30**, 275–282 (2009)
11. Bly, A.G., Woodard, H.J.: Foliar nitrogen application timing influence on grain yield and protein concentration of hard red winter and spring wheat. Agron. J. **95**, 335–338 (2003)
12. Boote, K.J., Gallaher, R.N., Robertson, W.K., Hinson, K., Hammond, L.C.: Effect of foliar fertilization on photosynthesis, leaf nutrition, and yield of soybeans. Agron. J. **70**, 787–791 (1978)
13. Bouma, D.: The response of subterranean clover (*Trifolium subterraneum* L.) to foliar applications of phosphorus. Aust. J. Agric. Res. **20**, 435–445 (1969)
14. Cakmak, I.: Enrichment of cereal grains with zinc: agronomic or genetic biofortification? Plant Soil **302**, 1–17 (2008)
15. Chaudhury, J., Mandal, U.K., Sharma, K.L., Ghosh, H., Mandal, B.: Assessing soil quality under long-term rice-based cropping system. Commun. Soil Sci. Plan. **36**, 1141–1161 (2005)
16. Choudhary, O.P., Singh, Y.: Nutrient management in irrigated agriculture: issues and strategies. In: FAI Annual Seminar 2018 Preprint, pp. 34–56. FAI, New Delhi (2018)
17. Datta, M., Singh, N.P.: Nutrient management in rice-based cropping systems as influenced by applying cattle manure alone or in combination with fertilizers in upland acid soils of Tripura. J. Ind. Soc. Soil Sci. **58**, 94–98 (2010)
18. Dey, P.: Soil test crop response: what can be learnt? In: Jat, M.L., et al. (eds.) National Dialogue on Efficient Nutrient Management for Improving Soil Health. Extended Summaries, pp. 25–31. TAAS, ICAR, CIMMYT, IPNI, CSISA, FAI, New Delhi, India (2015)
19. Dey, P.: AICRP on soil test crop response-research achievements and future directions. Indian J. Fertil. **15**(4), 376–385 (2019)

20. Dey, P., Jain, J.M.: Prediction of yield and nitrogen uptake of wheat from net nitrogen balance after rice in rice–wheat cropping system. Commun. Soil Sci. Plan. **44**(15), 2288–2298 (2013)
21. Dey, P., Sarkar, A.K.: Revisiting indigenous farming knowledge of Jharkhand (India) for conservation of natural resources and combating climate change. Indian J. Tradit. Knowl. **10**(1), 71–79 (2011)
22. Dobermann, A., Witt, C.: The evolution of site-specific nutrient management in irrigated rice systems of Asia. In: Dobermann, A., Witt, C., Dawe, D. (eds.) Increasing Productivity of Intensive Rice Systems Through Site-Specific Nutrient Management, pp. 76–100. IRRI (International Rice Research Institute), Science Publishers Inc. (2004)
23. Dobermann, A., Cassman, K.G., Peng, S., Tan, P.S., Phung, C.V., Sta Cruz, P.C., Bajita, J.B., Adviento, M.A.A., Olk, D.C.: Precision nutrient management in intensive irrigated rice systems. In: Proceedings of the International Symposium on Maximizing Sustainable Rice Yields Through Improved Soil and Environmental Management, Khon Kaen, Thailand, pp. 133–154 (1996)
24. Dobermann, A., Witt, C., Dawe, D., Abdulrachman, S., Gines, H.C., Nagarajan, R., Satawathananont, S., Son, T.T., Tan, P.S., Wang, G.H., Chien, N.V.: Site-specific nutrient management for intensive rice cropping systems in Asia. Field Crops Res. **74**, 37–66 (2002)
25. Erenoglu, B., Nikolic, M., Romheld, V., Cakmak, I.: Uptake and transport of foliar applied zinc (^{65}Zn) in bread and durum wheat cultivars differing in zinc efficiency. Plant Soil **241**, 251–257 (2002)
26. FAO: World Fertilizer Trends and Outlook to 2020. FAO, Rome (2018)
27. FCO: The Fertiliser Control Order 1985, 15th edn., 322 pp. The Fertiliser Association of India, New Delhi (2015)
28. Fowler, D., Coyle, M., Skiba, U., Sutton, M.A., Cape, J.N., Reis, S., Sheppard, .L.J., Jenkins, A., Grizzetti, B., Galloway, J.N., Vitousek, P.: The global nitrogen cycle in the twenty-first century. Philos. Trans. R. Soc. Lond. B Biol. Sci. **368**(1621), 2013.0164 (2013)
29. Garcia, R.L., Hanway, J.J.: Foliar fertilization of soybeans during seed-filling period. Agron. J. **68**, 653–657 (1976)
30. Geisseler, D., Scow, K.M.: Long-term effects of mineral fertilizers on soil microorganisms—a review. Soil Biol. Biochem. **75**, 54–63 (2014)
31. Ghai, S.K., Rao, D.L.N., Batra, L.: Nitrogen contribution to wetland rice by green manuring with *Sesbania* spp. in an alkali soil. Biol. Fertil. Soils **6**, 22–25 (1988)
32. Gholami, A., Akhlaghi, S., Shahsavani, S., Farrokhi, N.: Effects of urea foliar application on grain yield and quality of winter wheat. Commun. Soil Sci. Plant Anal. **42**, 719–727 (2011)
33. Girma, K., Martin, K.L., Freeman, K.W., Mosali, J., Teal, R.K., Raun, W.R., Moges, S.M., Arnall, B.: Determination of the optimum rate and growth stage for foliar applied phosphorus in corn. Commun. Soil Sci. Plan. **38**, 1137–1154 (2007)
34. Guo, J.H., Liu, X.J., Zhang, Y., Shen, J.L., Han, W.X., Zhang, W.F., Christie, P., Goulding, K.W.T., Vitousek, P.M., Zhang, F.S.: Significant acidification in major Chinese croplands. Science **327**, 1008–1010 (2010)
35. Haque, M.A., Jahiruddin, M., Islam, M.S., Rahman, M.M., Saleque, M.A.: Effect of bioslurry on the yield of wheat and rice in the wheat-rice cropping system. Agric. Res. **7**, 432–442 (2018)
36. Hatfield, J.L., Gitelson, A.A., Schepers, J.S., Walthall, C.L.: Application of spectral remote sensing for agronomic decisions. Agron. J. **100**(Supplement 3), S-117 (2008)
37. Hedley, C.: The role of precision agriculture for improved nutrient management on farms. J. Sci. Food Agric. **95**(1), 12–19 (2015)
38. Hedley, M.J., McLaughlin, M.J.: Reactions of phosphate fertilizers and by-products in soils. In: Sharpley, A.N. (ed.) Phosphorus: Agriculture and the Environment, pp. 181–252. ASA, CSSA, SSSA, Madison, WI (2005)
39. Hegde, D.M., Dwivedi, B.S.: Nutrient management in rice-wheat cropping system in India. Fertil. News **37**(2), 27–41 (1992)
40. IRRI: Use of Leaf Color Chart (LCC) for N Management in Rice. IRRI, Phillipines (1996)
41. Janssen, B.H., Guiking, F.C.T., Van der Eijk, D., Smaling, E.M.A., Wolf, J., van Reuler, H.: A system for quantitative evaluation of the fertility of tropical soils (QUEFTS). Geoderma **46**, 299–318 (1990)

42. Jinghua, G.: Synchrotron radiation, soft X-ray spectroscopy and nano-materials. J. Nanotechnol. **1**, 1–21 (2004)
43. Krapp, A.: Plant nitrogen assimilation and its regulation: a complex puzzle with missing pieces. Curr. Opin. Plant Biol. **25**, 115–122 (2015)
44. Kumar, S., Sharma, I.P., Raina, J.N.: Effect of levels and application methods of irrigation and mulch materials on strawberry production in north-west Himalayas. J. Indian Soc. Soil Sci. **53**, 60–65 (2005)
45. Lacroux, F., Tregoat, O., Van Leeuwen, C., Pons, A., Tominaga, T., Valérie, L.-C., Dubourdieu, D.: Effect of foliar nitrogen and sulphur application on aromatic expression of *Vitis vinifera* L. cv. Sauvignon blanc. J. Int. Sci. Vigne Vin. **42**(3), 125–132 (2008)
46. Ladha, J.K., Reddy, C.K., Padre, A.T., van Kessel, C.: Role of nitrogen fertilization in sustaining organic matter in cultivated soils. J. Environ. Qual. **40**, 1756–1766 (2011)
47. Lewis, B.: In: Drip Irrigation for Row Crops, Cooperative Extension Service Circular 573, p 24. New Mexico State University, Las Cruces (2001)
48. Li, S., Tian, Y., Wu, K., Ye, Y., Yu, J., Zhang, J., Liu, Q., Hu, M., Li, H., Tong, Y., Harberd, N.P.: Modulating plant growth–metabolism coordination for sustainable agriculture. Nature **560**(7720), 595–600 (2018)
49. Ling, G., Bextine, B.: Precising farming increases crop yields. Sci. Am. **317**, 34–35 (2017)
50. Linquist, B.A., Liu, L., van Kessel, C., van Groenigen, K.J.: Enhanced efficiency nitrogen fertilizers for rice systems: meta-analysis of yield and nitrogen uptake. Field Crops Res. **154**, 246–254 (2013)
51. Liu, R., Lal, R.: Synthetic apatite nano-particles as a phosphorous fertilizer for soybean (*Glycine max*). Sci. Rep. **4**, 5686–5691 (2014)
52. Liu, X., Feng, Z., Zhang, S., Zhang, J., Xiao, Q., Wang, Y.: Preparation and testing of cementing nano-subnano composites of slow- or controlled release of fertilizers. Sci. Agric. Sin. **39**, 1598–1604 (2006)
53. Lori, M., Symnaczik, S., Mäder, P., De Deyn, G., Gattinger, A.: Organic farming enhances soil microbial abundance and activity—a meta-analysis and meta-regression. PLoS One **12**(7) (2017)
54. Lu, C., Tian, H.: Global nitrogen and phosphorus fertilizer use for agriculture production in the past half century: shifted hot spots and nutrient imbalance. Earth Syst. Sci. Data **9**, 181–192 (2017)
55. Mader, P., Fließbach, A., Dubois, D., Gunst, L., Fried, P., Niggli, U.: Soil fertility and biodiversity in organic farming. Science **296**, 1694–1697 (2002)
56. Magen, H.: An overview of some practical aspects. Fertil. News **40**, 97–100 (1995)
57. Millar, N., Urrea, A., Kahmark, K., Shcherbak, I., Robertson, G.P., Ortiz-Monasterio, I.: Nitrous oxide (N_2O) flux responds exponentially to nitrogen fertilizer in irrigated wheat in the Yaqui Valley, Mexico. Agric. Ecosyst. Environ. **261**, 125–132 (2018)
58. Molin, J.P., Tavares, T.R.: Sensor systems for mapping soil fertility attributes: challenges, advances, and perspectives in Brazilian tropical soils. Engenharia Agric. **39**(SPE), 126–147 (2019)
59. Mosali, J., Desta, K., Teal, R.K., Freeman, K.W., Martin, K.L., Lawles, J.W., Raun, W.R.: Effect of foliar application of phosphorus on winter wheat grain yield, phosphorus uptake, and use efficiency. J. Plant Nutr. **29**, 2147–2163 (2006)
60. Mukhopadhyay, S.S.: Nanotechnology in agriculture: prospects and constraints. Nanotechnol. Sci. Appl. **7**, 63–71 (2014)
61. Muralidharudu, Y., Subba Rao, A., Sammi Reddy, K. (eds.): District-Wise Soil Test Based Fertilizer and Manure Recommendations for Balanced Nutrition of Crops, p. 270. Indian Institute of Soil Science, Bhopal (2012)
62. Mustaffa, M.M., Kumar, V.: Banana production and productivity enhancement through spatial, water and nutrient management. J. Hort. Sci. **7**(1), 1–28 (2012)
63. Nanda, R.S.: Fertigation to enhance farm productivity. Ind. J. Fertil. **6**(2), 13–22 (2010)
64. Narayanamoorthy, A.: Beneficial impact of drip irrigation. A study based on western India. Water Resour. J. **195**, 17–25 (1997)

65. NCAT, Oregon Tilth: Nutrient Management Plan (590) for Organic Systems. Western State Implementation Guide, p. 30. National Centre for Appropriate Technology (NCAT) and Oregon Tilth, USA (2014)
66. Noack, S.R., McBeath, T.M., McLaughlin, M.J.: Potential for foliar phosphorus fertilisation of dryland cereal crops: a review. Crop Pasture Sci. **61**, 659–669 (2010)
67. Pathak, H.: Greenhouse gas emission from Indian agriculture: trends, drivers and mitigation strategies. Proc. Ind. Natl. Sci. Acad. **81**, 1133–1149 (2015)
68. Praveena Katharine, S., Santhi, R., Chellamuthu, S., Ravikumar, V., Natesan, R., Dey, P.: Optimisation of fertiliser doses through inductive cum targeted yield model for transgenic cotton under drip fertigation on an Inceptisol. Mad. Agric. J. **101**(1–3), 33–39 (2014)
69. Qurashi, T.A., Salam, M.A., Jannat, M., Rabbani, M.G.: Evaluation of urea super granule as a source of nitrogen in transplant Aman rice. Progr. Agric. **24**(1–2), 29–37 (2013)
70. Rabani-Foroutagheh, M., Hamidoghli, Y., Mohajeri, S.A.: Effect of split foliar fertilisation on the quality and quantity of active constituents in saffron (*Crocus sativus* L.). J. Sci. Food Agric. **94**(9), 1872–1878 (2014)
71. Ramamurthy, V., Naidu, L.G.K., Kumar, S.R., Srinivas, S., Hegde, R.: Soil-based fertilizer recommendations for precision farming. Curr. Sci. **97**, 641–647 (2009)
72. Rao, D.L.N.: Slow-release urea fertilizers—effect on floodwater chemistry, ammonia volatilization and rice growth in an alkali soil. Fertil. Res. **13**, 209–221 (1987)
73. Rao, D.L.N., Batra, L.: Ammonia volatilization from applied nitrogen in alkali soils. Plant Soil **70**, 219–228 (1983)
74. Rao, D.L.N., Gill, H.S.: Biomass production and nutrient recycling through litter by pigeon pea (*Cajanus cajan* L. Mill sp.). Bioresour. Technol. **54**, 123–128 (1995)
75. Rao, D.L.N., Gill, H.S.: Biomass and biofertilizer production by *Sesbania cannabina* in alkaline soils. Bioresour. Technol. **53**, 169–172 (1995)
76. Rawat, A.K., Rao, D.L.N., Sahu, R.K.: Effect of soybean inoculation with *Bradyrhizobium* and wheat inoculation with *Azotobacter* on their productivity and N turnover in a Vertisol. Arch. Agron. Soil Sci. **59**, 1559–1571 (2013)
77. Reddy, D.D., Subba Rao, A., Sammi Reddy, K., Takkar, P.N.: Yield sustainability and phosphorus utilization in soybean-wheat system on Vertisols in response to integrated use of manure and fertilizer phosphorus. Field Crops Res. **62**, 180–190 (1999)
78. Reddy, K.S., Mohanty, M., Rao, D.L.N., Subba Rao, A., Pandey, M., Singh, M., Dixit, S.K., Dalal, R.C., Blamey, F.P.C., Menzies, N.W.: Farmer involvement in the development and adoption of improved nutrient management technologies using the mother-baby trial approach in Vertisols. Proc. Ind. Natl. Sci. Acad. India Sect. B Biol. Sci. **85**(1), 51–62 (2015). https://doi.org/10.1007/s40011-013-024-1
79. Reddy, K.S., Prabhakar, M., Sharma, K.L., Rama Rao, C.A., Vanaja, M., Srinivasa Rao, M., Gopinath, K.A., Sarkar, B., Ramana, D.B.V., Subba Rao, A.V.M., Sarath Chandran, M.A., Pankaj, P.K., Prasad, J.V.N.S., Bhaskar, S., Alagusundaram, K.: Research Highlights, p. 128. National Innovations in Climate Resilient Agriculture (NICRA), Central Research Institute for Dryland Agriculture, Hyderabad (2018)
80. Sammi Reddy, K.: Dryland agriculture in India: status and the way forward. Andhra Agric. J. **65**(3), 504–518 (2019)
81. Sammi Reddy, K., Mohanty, M., Rao, D.L.N., Singh, M., Dalal, R.C., Subba Rao, A., Pandey, M., Menzies, N.: Nitrogen mineralization in a Vertisol from organic manures, green manures and crop residues in relation to their quality. Agrochimica (Italy) **LII**(6), 377–388 (2008)
82. Sapkota, T.B., Majumdar, K., Jat, M.L., Kumar, A., Bishnoi, D.K., McDonald, A.J., Pampolino, M.: Precision nutrient management in conservation agriculture based wheat production of Northwest India: profitability, nutrient use efficiency and environmental footprint. Field Crops Res. **155**, 233–244 (2014)
83. Sharma, L.K., Bali, S.K.: A review of methods to improve nitrogen use efficiency in agriculture. Sustainability **10**(1), 1–23 (2017)
84. Shashidhara, K.K., Bheemappa, A., Hirevekanagoudar, L.V., Shashidhar, K.C.: Benefits and constraints in adoption of drip irrigation among the plantation crop growers. Karnataka J. Agric. Sci. **20**(1), 82–84 (2007)

85. Shukla, A.K., Ladha, J.K., Singh, V.K., Dwivedi, B.S., Balasubramanian, V., Gupta, R.K., Sharma, S.K., Singh, Y., Pathak, H., Pandey, P.S., Padre, A.T.: Calibrating the leaf color chart for nitrogen management in different genotypes of rice and wheat in a systems perspective. Agron. J. **96**, 1606–1621 (2004)
86. Singh, B.: Fertilizer management for sustenance of soil health. Ind. J. Fertil. **11**, 63–72 (2015)
87. Singh, M., Wanjari, R.H.: Relative contribution of major nutrients in crop productivity under long term fertilizer experiments in India. Ind. J. Fertil. **14**(7), 28–33 (2018)
88. Singh, B., Singh, V., Singh, Y., Kumar, A., Sharma, S., Thind, H.S., Choudhary, O.P., Vashistha, M.: Site-specific fertilizer nitrogen management in irrigated wheat using chlorophyll meter (SPAD meter) in the north-western India. J. Ind. Soc. Soil Sci. **66**, 53–65 (2018)
89. Smith, K.A., Chambers, B.J.: Utilizing the nitrogen content of organic manures on farms—problems and practical solutions. Soil Use Manage. **9**, 105–111 (1993)
90. Sneh, M.: The history of fertigation in Israel. Int. Water. Irrig. Rev. **15**, 17–20 (1995)
91. Soman, P.: Improving water use efficiency to enhance crop productivity. In: FAI Annual Seminar Proceedings, New Delhi (2009)
92. Subramanian, K.S., Sharmil Rahale, C.: Nano-fertilizer formulations for balanced fertilization of crops. Paper presented at ISSS Conference, New Delhi, 21–25 Dec 2009
93. Swarbreck, S.M., Wang, M., Wang, Y., Kindred, D., Sylvester-Bradley, R., Shi, W., Bentley, A.R., Griffiths, H.: A roadmap for lowering crop nitrogen requirement. Trends Plant Sci. **24**, 892–904 (2019)
94. Swarup, A.: Lessons from long-term fertilizer experiments in improving fertilizer use efficiency and crop yields. Fertil. News **47**(12), 59–73 (2002)
95. Tandon, H.L.S.: Fertiliser research some unanswered questions. Ind. J. Fertil. **8**(7), 26–31 (2012)
96. Tegeder, M., Masclaux-Daubresse, C.: Source and sink mechanisms of nitrogen transport and use. New Phytol. **217**, 35–53 (2017)
97. Teixeira, L.A.J., Quaggio, J.A., Mellis, E.V.: Enhancing nutrient use efficiency in banana due to irrigation and fertigation. Rev. Bras. Fruticult. **33**, 272–278 (2011)
98. Ullah, H., Santiago-Arenas, R., Ferdous, Z., Attia, A., Datta, A.: Improving water use efficiency, nitrogen use efficiency, and radiation use efficiency in field crops under drought stress: a review. Adv. Agron. **156**, 109–157 (2019)
99. Varga, B., Svecnjak, Z.: The effect of late-season urea spraying on grain yield and quality of winter wheat cultivars under low and high basal nitrogen fertilization. Field Crops Res. **96**, 125–132 (2006)
100. Varinderpal-Singh, Singh, B., Singh, Y., Thind, H.S., Buttar, G.S., Kaur, S., Singh, M., Kaur, S., Bhowmik, A.: Site-specific fertilizer nitrogen management for timely sown irrigated wheat (*Triticum aestivum* L. and *Triticum turgidum* L. ssp. durum) genotypes. Nutr. Cycl. Agroecosyst. **109**, 1–16 (2017)
101. Wander, M.M., Drinkwater, L.E.: Fostering soil stewardship through soil quality assessment. Field Crops Res. **15**, 61–73 (2000)
102. Willer, H., Lernoud, J., Kemper, L.: The world of organic agriculture 2018: summary. In: The World of Organic Agriculture. Statistics and Emerging Trends 2018, pp. 22–31. Research Institute of Organic Agriculture FiBL and IFOAM-Organics International (2018)
103. Witt, C., Dobermann, A., Abdulrachman, S., Gines, H.C., Wang, G.H., Nagarajan, R., Satawathananont, S., Son, T.T., Tan, P.S., Tiem, L.V., Simbahan, G.C., Olk, D.C.: Internal nutrient efficiencies in irrigated lowland rice in tropical and subtropical Asia. Field Crops Res. **63**, 113–138 (1999)
104. Wortman, S.E., Wortmann, C.S., Pine, A.L., Shapiro, C.A., Thompson, A.A., Little, R.S.: Nutrient management in organic farming. In: G2295: Index Crops, Crop Production/Field Crops. Nebraska Extension (2017)
105. Yang, Y.C., Zhang, M., Li, Y., Fan, X.H., Geng, Y.Q.: Improving the quality of polymer-coated urea with recycled plastic, proper additives, and large tablets. J. Agric. Food Chem. **60**(45), 11229–11237 (2012)

106. Zhang, J.J., He, P., Xu, X.P., Wang, Y.L., Jia, L.L., Cui, R.Z., Wang, H.T., Zhao, S.C., Ullah, S.: Nutrient expert improves nitrogen efficiency and environmental benefits for summer maize in China. Agron. J. **109**, 1082–1090 (2017)

Index

G. Benckiser (ed.), *Soil and Recycling Management in the Anthropocene Era*, Environmental Science and Engineering,
https://doi.org/10.1007/978-3-030-51886-8

The manufacturer's authorised representative in the EU is Springer Nature Customer Service Centre GmbH, Europaplatz 3, 69115 Heidelberg, Germany. If you have any concerns regarding our products, please contact ProductSafety@springernature.com

Printed and bound by CPI Group (UK) Ltd, Croydon, CR0 4YY
30/09/2026
02225665-0015